Practical Engineering, Process, and Reliability Statistics

Also available from ASQ Quality Press:

The Certified Pharmaceutical GMP Professional Handbook
ASQ FD&C Division and Mark A. Durivage, editor

Reliability Data Analysis with Excel and Minitab
Kenneth S. Stephens

Zero Acceptance Number Sampling Plans, Fifth Edition
Nicholas L. Squeglia

The Certified Quality Engineer Handbook, Third Edition
Connie M. Borror, editor

The Quality Toolbox, Second Edition
Nancy R. Tague

Root Cause Analysis: Simplified Tools and Techniques, Second Edition
Bjørn Andersen and Tom Fagerhaug

The Certified Six Sigma Green Belt Handbook, Second Edition
Roderick A. Munro, Govindarajan Ramu, and Daniel J. Zrymiak

The Certified Manager of Quality/Organizational Excellence Handbook, Fourth Edition
Russell T. Westcott, editor

The Certified Six Sigma Black Belt Handbook, Second Edition
T.M. Kubiak and Donald W. Benbow

The ASQ Auditing Handbook, Fourth Edition
J.P. Russell, editor

The ASQ Quality Improvement Pocket Guide: Basic History, Concepts, Tools, and Relationships
Grace L. Duffy, editor

To request a complimentary catalog of ASQ Quality Press publications, call 800-248-1946, or visit our website at http://www.asq.org/quality-press.

Practical Engineering, Process, and Reliability Statistics

Mark Allen Durivage

ASQ Quality Press
Milwaukee, Wisconsin

American Society for Quality, Quality Press, Milwaukee 53203
© 2015 by ASQ
All rights reserved. Published 2014
Printed in the United States of America
20 19 18 17 16   5 4 3

Library of Congress Cataloging-in-Publication Data

Durivage, Mark Allen, author.
   Practical engineering, process, and reliability statistics / Mark Allen Durivage.
      pages cm
   Includes bibliographical references and index.
   ISBN 978-0-87389-889-8 (hard cover : alk. paper)
   1. Engineering—Statistical methods. 2. Reliability (Engineering)—Statistical methods.
 I. Title.

   TA340.D87 2014
   620'.0045—dc23   2014030697

ISBN: 978-0-87389-889-8

Publisher: Lynelle Korte
Acquisitions Editor: Matt T. Meinholz
Project Editor: Paul Daniel O'Mara
Production Administrator: Randall Benson

ASQ Mission: The American Society for Quality advances individual, organizational, and community excellence worldwide through learning, quality improvement, and knowledge exchange.

Attention Bookstores, Wholesalers, Schools, and Corporations: ASQ Quality Press books, video, audio, and software are available at quantity discounts with bulk purchases for business, educational, or instructional use. For information, please contact ASQ Quality Press at 800-248-1946, or write to ASQ Quality Press, P.O. Box 3005, Milwaukee, WI 53201-3005.

To place orders or to request ASQ membership information, call 800-248-1946. Visit our website at http://www.asq.org/quality-press.

 Printed on acid-free paper

Quality Press
600 N. Plankinton Ave.
Milwaukee, WI 53203-2914
E-mail: authors@asq.org
ASQ The Global Voice of Quality™

Table of Contents

List of Figures and Tables

Preface

Practical Engineering, Process, and Reliability Statistics was written to aid quality technicians and engineers. It is a compilation of 30 years of quality related work experience. I am frustrated by the number of books necessary, at times, to provide statistical support. To that end, the intent of this book is to provide the quality professional working in virtually any industry a quick, convenient, and comprehensive guide to properly utilizing statistics in an efficient and effective manner.

This book assumes the reader has been exposed to basic statistical concepts. All examples utilize algebra-based problem-solving techniques rather than calculus.

This book will be a useful reference when preparing for and taking many of the ASQ quality certification examinations, including the Certified Quality Technician (CQT), Certified Six Sigma Green Belt (CSSGB), Certified Quality Engineer (CQE), Certified Six Sigma Black Belt (CSSBB), and Certified Reliability Engineer (CRE).

Acknowledgments

I would like to acknowledge the previous work of Robert A. Dovich in *Quality Engineering Statistics* and *Reliability Statistics*. This book is an expansion of his efforts in an attempt to continue the method of presenting statistical applications in a simple, easy-to-follow format. I would like to thank those who have inspired, taught, and trained me throughout my academic and professional career. Additionally, I would like to thank ASQ Quality Press, especially Matt Meinholz, Acquisitions Editor, and Paul O'Mara, Managing Editor, for their expertise and technical competence, which made this project a reality, and my friend from the ASQ Reliabity Division Marc Banghart for reviewing this book. Lastly, I would like to acknowledge the patience of my wife Dawn and my sons Jack and Sam, which allowed me time to research and write *Practical Engineering, Process, and Reliability Statistics.*

Limit of Liability/
Disclaimer of Warranty

The author has put forth his best efforts in compiling the content of this book; however, no warranty with respect to the material's accuracy or completeness is made. Additionally, no warranty is made in regards to applying the recommendations made in this book to any business structure or environments. Businesses should consult regulatory, quality, and/or legal professionals prior to deciding on the appropriateness of advice and recommendations made within this book. The author shall not be held liable for loss of profit or other commercial damages resulting from the employment of recommendations made within this book including special, incidental, consequential, or other damages.

1

Point Estimates and Measures of Dispersion

When performing statistical tests, we usually work with data that are samples drawn from a population. We use sample data to make estimates about the population. The first estimate is usually a point estimate (central tendency).

As we will see in Chapter 2, Confidence Intervals, these point estimates are subject to sampling error and should be interpreted with caution, especially for small sample sizes. The accuracy of the point estimates becomes higher as the sample size gets larger.

There are several point estimates commonly made by quality technicians, quality engineers, and reliability engineers. These include estimates of central tendency, such as the mean (average), median, and mode.

Estimates of dispersion include the range, variance, standard deviation, coefficient of variation, and others. Descriptions of the shape of the distribution include skewness and kurtosis.

1.1 ESTIMATES OF CENTRAL TENDENCY FOR VARIABLES DATA

The most common measure of central tendency is the *average* or *sample mean*. The true (unknown) population mean is denoted by the letter μ, and is estimated by $\bar{X}$ (X-bar). To estimate the parameter μ using $\bar{X}$ we use the following formula:

$$\bar{X} = \frac{\Sigma X}{n}$$

where

X = Data point

n = Number of data points

Example: Using the following seven data points, estimate the population mean μ by finding $\bar{X}$:

43, 45, 40, 39, 42, 44, 41.

$$\bar{X} = \frac{\Sigma X}{n} = \frac{43+45+40+39+42+44+41}{7} = \frac{294}{7} = 42$$

Another estimate of central tendency or location is the *median*. The median is a simpler value to determine because it can be determined without a mathematical calculation. The median value is most useful when there are outlying data points that could artificially inflate or deflate the arithmetic mean. To find the median, place the data points in an ordered form, generally the lowest value on the left and the greatest value on the right.

Example: Using the following seven data points, determine the median:

$$43, 45, 40, 39, 42, 44, 41$$

Order the data points and select the point in the middle:

$$39, 40, 41, \textbf{42}, 43, 44, 45$$

This example yields 42 as the median value for these seven data points. In the case where there is an even number of data points, add the two values in the middle and divide by two.

Example: Using the following six data points, determine the median:

$$43, 40, 39, 42, 44, 41.$$

Order the data points and select the points in the middle, and divide by two:

$$39, 40, \textbf{41}, \textbf{42}, 43, 44$$

$$\frac{41+42}{2} = 41.5$$

The calculated median value is 41.5.

The *mode* is the most frequently occurring value(s) in a set of data. A set of data may contain one mode, two modes (bimodal), many modes, or no mode.

Example: Given the following data points, determine the mode:

$$39, 40, \textbf{41}, \textbf{41}, 42, 43$$

The mode is 41, as it is the most frequently appearing value in the data set.

When the population set is unimodal and symmetrical, such as in the normal (Gaussian) distribution, the values of mean, median, and mode will occur at the same location, as shown in Figure 1.1. When the distribution is skewed, these values diverge, as shown in Figures 1.2 and 1.3

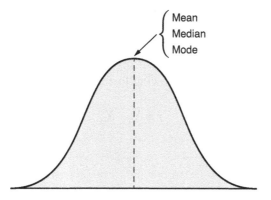

Figure 1.1   Normal distribution.

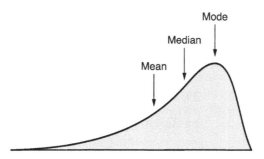

Figure 1.2   Negatively skewed distribution.

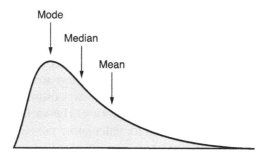

Figure 1.3   Positively skewed distribution.

1.2 RANGE FOR VARIABLES DATA

The range is the simplest method of measuring the spread of sample or population data. To calculate the range, we use the following formula:

$$r = X_h - X_l$$

where

r = Range

X_h = Largest value

X_l = Smallest value

Example: Using the following seven data points, determine the range:

$$43, 45, 40, 39, 42, 44, 41$$

Order the data points and select the largest and smallest values:

$$\mathbf{39}, 40, 41, 42, 43, 44, \mathbf{45}$$

$$r = X_h - X_l = 45 - 39 = 6$$

The range of this set of data is 6.

1.3 VARIANCE AND STANDARD DEVIATION FOR VARIABLES DATA

This section will focus on the two primary measures of dispersion or variation for individual values. The two principal measures of dispersion are the *variance* σ^2, which is estimated from sample data, and the statistic s^2, which is estimated by the statistic s. We see that the *standard deviation* is the square root of the variance. For variables data, when all values are available, the formula for calculating the population variance is

$$\sigma^2 = \frac{\Sigma(x - \mu)^2}{N}$$

where

μ = Population mean

x = Data point

N = The number of data points in the population

The formula for calculating the population standard deviation is

$$\sigma = \sqrt{\sigma^2} \quad \text{or} \quad \sigma = \sqrt{\frac{\Sigma(x-\mu)^2}{N}}$$

Example: Using the following seven data points, determine the population variance and standard deviation:

$$43, 45, 40, 39, 42, 44, 41$$

x	$(x-\mu)^2$
43	1
45	9
40	4
39	9
42	0
44	4
41	1
$\mu = 42$	28

$$\sigma^2 = \frac{\Sigma(x-\mu)^2}{N} = \frac{28}{7} = 4$$

The calculated variance is 4.

$$\sigma = \sqrt{\sigma^2} = \sqrt{4} = 2$$

The calculated standard deviation is 2.

Generally, though, the focus is on using a sample drawn from the population to make inferences about the population. To make an unbiased estimate of the population variance from sample data, the formula is

$$s^2 = \frac{\Sigma(X-\overline{X})^2}{n-1}$$

where

$\overline{X}$ = Sample mean

X = Data point

n = The number of data points in the sample

The formula for calculating the sample standard deviation is

$$s = \sqrt{s^2} \text{ or } s = \sqrt{\frac{\Sigma(X - \bar{X})^2}{n-1}}$$

Example: Using the following seven data points, determine the sample variance and standard deviation:

$$43, 45, 40, 39, 42, 44, 41$$

x	$(X - \bar{X})^2$
43	1
45	9
40	4
39	9
42	0
44	4
41	1
$\bar{X} = 42$	28 (Sum)

$$s^2 = \frac{\Sigma(X - \bar{X})^2}{n-1} = \frac{28}{7-1} = 4.67$$

The calculated variance is 4.67.

$$s = \sqrt{s^2} = \sqrt{4.67} = 2.16$$

The calculated standard deviation is 2.16.

The *coefficient of variation* is a normalized measurement of dispersion. The coefficient of variation is calculated by the following formula:

$$c_v = \frac{\sigma}{\mu} * 100 \text{ Population}$$

$$c_v = \frac{s}{\bar{X}} * 100 \text{ Sample}$$

where

μ = Population mean

$\bar{X}$ = Sample mean

σ = Population standard deviation

s = Sample standard deviation

Example: Calculate the coefficient of variation for the following population values:

$$\mu = 42 \text{ and } \sigma = 2$$

$$c_v = \frac{\sigma}{\mu} * 100 = \frac{2}{42} * 100 = 4.76\%$$

The calculated population coefficient of variation is 4.76%.

Example: Calculate the coefficient of variation for the following sample values:

$$\bar{X} = 42 \text{ and } s = 2$$

$$c_v = \frac{s}{\bar{X}} * 100 = \frac{2.16}{42} * 100 = 5.14\%$$

The calculated sample coefficient of variation is 5.14%.

1.4 SKEWNESS AND KURTOSIS FOR VARIABLES DATA

Skewness and kurtosis are two measures that describe the shape of the normal distribution. *Skewness* is a measure of symmetry about the center of the distribution. The skewness for a normal distribution should be zero. Negative skewness indicates the data are skewed to the left (the left tail is longer than the right tail) and positive skewness indicates the data are skewed to the right (the right tail is longer than the left tail). Please note that various sources indicate the use of other formulae. The formula to calculate skewness is

$$\text{Skewness} = \frac{\Sigma x (X - \bar{X})^3}{(n-1)s^3}$$

where

$\bar{X}$ = Sample mean

s = Sample standard deviation

X = Data point

n = The number of data points in the sample

Caution must be exercised when analyzing the calculated skewness from a sample of the population. The sample may not be indicative of the population. The following table provides a guideline for interpreting the calculated skewness value.

−1 or smaller	Highly skewed	1 or larger
−1 to −0.5	Moderately skewed	0.5 to 1
−0.5 to 0	Approximately symmetric	0 to 0.5

Kurtosis is a measure that describes the flatness or peakedness of a normal distribution. Generally, kurtosis describes the closeness of the data points relative to the center of the distribution. A normal distribution is called *mesokurtic*. A distribution that is flattened is *platykurtic*. A distribution that has a sharp peak is *leptokurtic*. Higher values indicate peakedness, and lower values indicate a less pronounced peak. The formula for calculating kurtosis is:

$$\text{Kurtosis} = \frac{\Sigma x(X - \bar{X})^4}{(n-1)s^4}$$

where

$\bar{X}$ = Sample mean

s = Sample standard deviation

X = Data point

n = The number of data points in the sample

Caution must be exercised when analyzing the calculated kurtosis from a sample of the population. The sample may not be indicative of the population. The following table provides a guideline for interpreting the calculated kurtosis value.

Kurtosis measures	
0	Mesokurtic
< 3	Platykurtic
> 3	Leptokurtic

Example: Using the following seven data points, determine the skewness and kurtosis:

43, 45, 40, 39, 42, 44, 41

X	$(X - \bar{X})^3$	$(X - \bar{X})^4$
43	1	1
45	27	81
40	−8	16
39	−27	81
42	0	0
44	8	16
41	−1	1
$\bar{X} = 42$		196 (Sum)
$s = 2.16$		

$$\text{Skewness} = \frac{\Sigma x (X - \bar{X})^3}{(n-1)s^3} = \frac{0}{(7-1)*2.16^3} = 0.00$$

$$\text{Kurtosis} = \frac{\Sigma x (X - \bar{X})^4}{(n-1)s^4} = \frac{196}{(7-1)*2.16^4} = 1.50$$

1.5 ESTIMATES OF CENTRAL TENDENCY FOR ATTRIBUTES DATA

When working with attributes data, measures of central tendency that are analogous to the average are the proportion or expected numbers of occurrences. For nonconforming product, the fraction nonconforming is estimated by the statistic p, where p is found as

$$p = \frac{\text{Number of nonconforming units}}{\text{Number of units tested}}$$

A more general equation could be written as

$$p = \frac{\text{Number of occurrences}}{\text{Total sample size}}$$

Example: If we test 700 units and find 16 to be nonconforming, the estimate of the population fraction nonconforming would be

$$p = \frac{16}{700} = 0.023$$

To convert to a percentage, multiply the *p* value by 100:

$$0.023 * 100 = 2.3\%$$

When there are several inspection points per unit, such as on a circuit board, there is a possibility of more than one nonconformance per unit. When the opportunity for occurrence is the same, as in identical units, the average number of nonconformances is estimated using the *c* statistic. The *c* statistic is calculated using the following formula:

$$c = \frac{\text{Number of nonconformances}}{\text{Number of items inspected}}$$

Example: If 500 circuit boards reveal 1270 nonconformances, the estimate of the number of nonconformances per unit in the population is

$$c = \frac{1270}{500} = 2.54 \text{ nonconformances per unit.}$$

1.6 ESTIMATES OF DISPERSION FOR ATTRIBUTES DATA

When working with fraction nonconforming, the estimate of the variance is a function of the binomial distribution and is given as

$$s^2 = p(1-p) \text{ Sample variance}$$

$$s = \sqrt{p(1-p)} \text{ Sample standard deviation}$$

$$s = \sqrt{s^2} \text{ Sample standard deviation}$$

where

   p = Fraction nonconforming

Example: If we test 700 units and find 16 to be nonconforming, the estimate of the population fraction nonconforming is *p* = 0.023. We calculate the variance as

$$s^2 = p(1-p) = 0.023(1-0.023) = 0.0225$$

The standard deviation is the square root of the variance:

$$s = \sqrt{s^2} = \sqrt{0.0225} = 0.15$$

If we are interested in the number of nonconformances per unit or similar count data, we may model the variance by the Poisson distribution. When the data can be modeled by the Poisson distribution, the variance is equal to the mean value.

Example: If 500 circuit boards reveal 1270 nonconformances, the estimate of the number of nonconformances per unit in the population is 2.54. Calculate the variance and standard deviation:

$$s^2 = 2.54$$

is the calculated sample variance.

$$s = \sqrt{2.54} = 1.594$$

is the calculated sample standard deviation.

2
Confidence Intervals

In Chapter 1, we made estimates from a single sample of data drawn from a population. If we were to take another sample from the population, it is likely that the estimates of the mean, variance, standard deviation, skewness, and kurtosis would be different, and none would be the exact value of the population.

To compensate for the sampling variation, we use the concept of *confidence intervals*. These intervals will contain the true but unknown population parameter for the percentage of the time chosen by the engineer. For example, if one were to calculate 95% confidence intervals for the mean, 95% of the intervals calculated from samples drawn from the population would contain the true population mean, while 5% would not.

2.1 CONFIDENCE INTERVAL FOR THE MEAN OF CONTINUOUS DATA

When the sample is large and the population standard deviation is known, or the distribution is normal, we may use the normal distribution to calculate the confidence interval for the mean. The formula is

$$\bar{X} \pm Z_{\alpha/2} \frac{\sigma}{\sqrt{n}}$$

where

$\bar{X}$ = The point estimate of the average

σ = The population standard deviation

n = The sample size

$Z_{\alpha/2}$ = The normal distribution value for a given confidence level (see Appendix D, Selected Double-Sided Normal Distribution Probability Points)

Example: If the mean of a sample drawn from a population is 2.15 and the population standard deviation is known to be 0.8, calculate the 95% confidence interval for the average if the sample size is 75.

From Appendix D, the value for $Z_{\alpha/2}$ is 1.960, and the 95% confidence interval is calculated as

$$\bar{X} \pm Z_{\alpha/2}\frac{\sigma}{\sqrt{n}} = 2.15 \pm 1.960 * \frac{0.8}{\sqrt{75}} = 2.15 \pm 0.181 = 1.969 \text{ to } 2.331$$

If the sample standard deviation is estimated from a relatively small (fewer than 100) sample, the formula becomes

$$\bar{X} \pm t_{\sigma/2,n-1} * \frac{s}{\sqrt{n}}$$

where

$\bar{X}$ = The point estimate of the average

s = The sample standard deviation

n = The sample size

$t_{\alpha/2,n-1}$ = Percentage points of the Student's t distribution value for a given confidence level for $(n - 1)$ degrees of freedom (df) (see Appendix E, Percentage Points of the Student's t-Distribution)

Example: If the mean of a sample drawn from a population is 2.15 and the standard deviation estimate is 0.8, calculate the 95% confidence interval for the mean if the sample size is 75.

From Appendix E, the value for $t_{\alpha/2}$ with 74 degrees of freedom is 2.000. Please note, the table did not have a value for 74 degrees of freedom, so therefore use the next smaller value, in this case 60.

$$\bar{X} \pm t_{\sigma/2,n-1} * \frac{s}{\sqrt{n}} = 2.15 \pm 2.000 * \frac{0.8}{\sqrt{75}} = 2.15 \pm 0.185 = 1.965 \text{ to } 2.335$$

2.2 CONFIDENCE INTERVAL FOR THE VARIANCE AND STANDARD DEVIATION

Unlike the confidence interval for the mean, the confidence interval for the variance will not be symmetrical about the point estimate since it is based on the chi-squared distribution. The formula for calculating the confidence interval for the variance is

$$\frac{(n-1)*s^2}{\chi^2_{\alpha/2,n-1}} \leq \sigma^2 \leq \frac{(n-1)*s^2}{\chi^2_{1-\alpha/2,n-1}}$$

The formula for calculating the confidence interval for the standard deviation is

$$\sqrt{\frac{(n-1)*s^2}{\chi^2_{\alpha/2,n-1}}} \leq \sigma \leq \sqrt{\frac{(n-1)*s^2}{\chi^2_{1-\alpha/2,n-1}}}$$

where

s = The sample standard deviation

s^2 = The sample variance

n = The sample size

$\chi^2_{\alpha/2,n-1}$ and $\chi^2_{1-\alpha/2,n-1}$ = Distribution of the chi-square value for a given confidence level for $(n-1)$ degrees of freedom (see Appendix F, Distribution of the Chi-Square)

Example: The sample variance (s^2) is estimated to be 0.673 from a sample size of 25. Calculate the 90% confidence interval for the population variance. From Appendix F, the appropriate chi-square value for $\alpha/2 = 0.05$ and 24 df is 36.415, and $1 - \alpha/2 = 0.95$ and 24 df is 13.848:

$$\frac{(n-1)*s^2}{\chi^2_{\alpha/2,n-1}} \leq \sigma^2 \leq \frac{(n-1)*s^2}{\chi^2_{1-\alpha/2,n-1}}$$

$$\frac{(25-1)*0.673}{36.415} \leq \sigma^2 \leq \frac{(25-1)*0.673}{13.848}$$

$$0.444 \leq \sigma^2 \leq 1.166$$

Based on the single sample of 25, we are 90% confident that the interval from 0.444 to 1.166 contains the true value of the variance.

Example: The sample standard deviation (s) is estimated to be 0.820 from a sample size of 25. Calculate the 90% confidence interval for the population variance. From Appendix F, the appropriate chi-square value for $\alpha/2 = 0.05$ and 24 df is 36.415, and for $1 - \alpha/2 = 0.95$ and 24 df is 13.848:

$$\sqrt{\frac{(n-1)*s^2}{\chi^2_{\alpha/2,n-1}}} \leq \sigma \leq \sqrt{\frac{(n-1)*s^2}{\chi^2_{1-\alpha/2,n-1}}}$$

$$\sqrt{\frac{(25-1)*0.820}{36.415}} \leq \sigma \leq \sqrt{\frac{(25-1)*0.820}{13.848}}$$

$$0.735 \leq \sigma \leq 1.192$$

Based on a sample of 25 pieces, we are 90% confident that the interval from 0.735 to 1.192 contains the true value of the standard deviation.

2.3 CONFIDENCE INTERVAL FOR THE FRACTION NONCONFORMING—NORMAL DISTRIBUTION

Due to the variability associated with point estimates, any estimates of the fraction non-conforming using the normal distribution are distributed as a second random variable, with the associated confidence interval estimates. To calculate the fraction nonconforming using the normal distribution, we use the equation

$$Z = \frac{x - \mu}{\sigma}$$

where

μ = The population mean

σ = The population standard deviation

x = Any point of interest, such as a specification limit

Z = The normal distribution value (see Appendix A, Normal Distribution Probability Points—Area below Z)

Example: The upper specification limit for a process is 90. The process average is 88.5, and the standard deviation is 1.23. Both of these values were estimated from a relatively large sample size of $n = 150$:

$$Z = \frac{x - \mu}{\sigma} = \frac{90 - 88.5}{1.23} = 1.220$$

Using Appendix A, Normal Distribution Probability Points—Area below Z, the Z value of 1.220 is 0.1112 or 11.12%.

In actuality, the estimates of the mean and standard deviation are point estimates. Using these values to estimate fraction nonconforming is intuitively dissatisfying. Instead, the following formulas (Weingarten 1982) can be used to calculate a confidence interval for the fraction nonconforming:

$$Z_{\text{UCL}} = Q - Z_{\alpha/2} * \sqrt{\frac{1}{n} + \frac{Q^2}{2n}}$$

and

$$Z_{\text{LCL}} = Q + Z_{\alpha/2} * \sqrt{\frac{1}{n} + \frac{Q^2}{2n}}$$

where

Z_{LCL} = The Z value for the upper confidence limit

Z_{UCL} = The Z value for the lower confidence limit

Q = The Z value calculated

n = The sample size

$Z_{\alpha/2}$ = The normal distribution value for a given confidence level (see Appendix D, Selected Double-Sided Normal Distribution Probability Points)

Example: Calculate the 95% confidence interval for the fraction nonconforming using the calculated Z value of 1.220. Please note that $Q = Z$. The sample size is 50.

Z_{α} = 1.960 (see Appendix D, Selected Double-Sided Normal Distribution Probability Points)

$$Z_{LCL} = Q - Z_{\alpha/2} * \sqrt{\frac{1}{n} + \frac{Q^2}{2n}} = 1.220 - 1.960 * \sqrt{\frac{1}{50} + \frac{1.220^2}{300}} = 1.009$$

and

$$Z_{UCL} = Q + Z_{\alpha/2} * \sqrt{\frac{1}{n} + \frac{Q^2}{2n}} = 1.220 + 1.960 * \sqrt{\frac{1}{50} + \frac{1.220^2}{300}} = 1.431$$

For $Z_{UCL} = 1.009$, referring to Appendix B (Normal Distribution Areas above Z), the estimated fraction nonconforming is equal to 0.1562 or 15.62%. For $Z_{LCL} = 1.431$, the estimated fraction nonconforming is equal to 0.0764 or 7.64%. Therefore, the 95% confidence limits on the fraction nonconforming are 0.0764 to 0.1562.

2.4 CONFIDENCE INTERVAL FOR PROPORTION (NORMAL APPROXIMATION OF THE BINOMIAL CONFIDENCE INTERVAL)

The point estimate for a proportion or fraction nonconforming is given in the following equation:

$$p = \frac{\text{Number of occurrences}}{\text{Total sample size}}$$

When the sample size is large, and $n(p)$ and $n(1 - p)$ are greater than or equal to 5, we can use the normal distribution to calculate the confidence interval for the proportion. The formula for the confidence interval for the proportion is

$$p \pm Z_{\alpha/2} * \sqrt{\frac{p(1-p)}{n}}$$

where

p = Proportion

n = The sample size

$Z_{\alpha/2}$ = The normal distribution value for a given confidence level (see Appendix D, Selected Double-Sided Normal Distribution Probability Points)

Example: Calculate the 90% confidence interval when we test 700 units and find 16 to be nonconforming, and the estimate of the population fraction nonconforming is 0.023:

$$p \pm Z_{\alpha/2} * \sqrt{\frac{p(1-p)}{n}}$$

$$0.023 \pm 1.645 * \sqrt{\frac{0.023 * 0.977}{700}} = 0.0093$$

$$0.023 \pm 0.0093 = 0.0137 \text{ to } 0.0323$$

The 90% confidence interval is the range 0.0137 to 0.0323.

Notes:

Accuracy suffers when $np < 5$ or $n(1 - 9) < 5$

Calculation is not possible when $p = 0$ or 1

2.5 SMALL SAMPLE SIZE CONFIDENCE INTERVALS

When the sample size is medium to small, the requirements to use the equations in Section 2.4 may not be satisfied. In these cases, we may use another set of equations to calculate the confidence interval for the proportion of occurrences in a sample. These equations, adapted from Burr (1974), are suitable with some minor changes in notations made for continuity. The confidence intervals are calculated by the following formulas:

$$\Phi_L = \frac{2r - 1 + Z_{\alpha/2}^2 - Z_{\alpha/2} * \sqrt{[(2r-1)(2n-2r+1)/n] + Z_{\alpha/2}^2}}{2(n + Z_{\alpha/2}^2)}$$

$$\Phi_U = \frac{2r+1+Z_{\alpha/2}^2 + Z_{\alpha/2} * \sqrt{[(2r+1)(2n-2r-1)/n] + Z_{\alpha/2}^2}}{2(n+Z_{\alpha/2}^2)}$$

where

r = Number of occurrences in the sample

n = The sample size

Z_α = The normal distribution value for a given confidence level (see Appendix D, Selected Double-Sided Normal Distribution Probability Points)

Example: In a sample of 50 amplifiers, seven are found to be nonconforming. Calculate the 95% confidence limits for the true fraction nonconforming:

$$\Phi_L = \frac{2*7-1+1.960^2 -1.960*\sqrt{[(2*7-1)(2*50-2*7+1)/50]+1.960^2}}{2(50+1.96^2)} = 0.063$$

$$\Phi_U = \frac{2*7+1+1.960^2 +1.96*\sqrt{[(2*7+1)(2*50-2*7-1)/50]+1.96^2}}{2(50+1.96^2)} = 0.274$$

The 95% confidence interval for the fraction nonconforming is 0.063 to 0.274. This is compared to the single point estimate of 7/50 = 0.14.

2.6 CONFIDENCE INTERVAL FOR THE POISSON DISTRIBUTED DATA

To calculate the confidence interval for Poisson distributed data, the relationship given by Nelson between the chi-squared distribution and the Poisson distribution allows for simple calculations of Poisson confidence intervals. To calculate the upper confidence limit for the number of occurrences, calculate the appropriate degrees of freedom for the chi-squared table as $v = 2(r + 1)$. To calculate the lower confidence limit for the number of occurrences, calculate the df as $2r$.

When calculating 90% two-sided confidence intervals, use the columns labeled 0.05 and 0.95, as well as the appropriate df for upper and lower confidence limits. The values obtained in Appendix F, Distribution of the Chi-Square, are divided by two for the required estimate.

Example: An examination of a complex assembly noted 13 nonconformances. Find the 90% confidence interval for the number of nonconformances:

Upper confidence limit df = 2(13 + 1) = 28. The chi-squared table value for 28 df with an α of 0.05 is 41.337. The upper confidence limit is 41.337/2 = 20.67.

Lower confidence limit df = 2(13) = 26. The chi-squared table value for 26 df with an α of 0.95 is 15.379. The upper confidence limit is 15.379/2 = 7.69.

Thus, the 90% confidence interval for the number of nonconformances is 7.69 to 20.67.

3

Tolerance and Prediction Intervals

Confidence intervals are estimates of population parameters. We will now introduce the concept of tolerance and prediction intervals.

3.1 TOLERANCE INTERVALS

Tolerance intervals are used to estimate where a proportion of a population is located. Tolerance intervals are given with two percentage values. The first percentage value is the degree of confidence that the interval contains a certain percentage of each individual measurement in the population. The second percentage value is the fraction of the population the interval contains. Tolerance intervals assume a normally distributed population. Tolerance intervals can be calculated for unilateral (one-sided) or bilateral (two-sided) values using the following formulas:

$$\bar{X} \pm K_2 s \quad \text{Two-sided bilateral limits}$$

$$X_L = \bar{X} - K_1 s \quad \text{One-sided unilateral lower limit}$$

$$X_U = \bar{X} + K_1 s \quad \text{One-sided unilateral upper limit}$$

where

K_2 = Two-sided factor (see Appendix H, Tolerance Interval Factors)

K_1 = One-sided factor (see Appendix H, Tolerance Interval Factors)

s = Sample standard deviation

$\bar{X}$ = The sample mean

X_L and X_U = Lower and upper target values

Example: A process has an average of 1.500 and a sample standard deviation of 0.002. If 500 parts are produced, what is the tolerance interval that we can be 95% confident will contain 99% of the products produced?

$$\bar{X} \pm K_2 s$$

$$1.500 \pm 2.721 * 0.002 = 0.0054$$

$$1.500 \pm 0.0054 = 1.4946 \text{ to } 1.5054$$

We can be 95% confident that 99% of the parts are between 1.4946 to 1.5054.

Example: A process has an average of 1.500 and a sample standard deviation of 0.002. If 80 parts are produced, what is the tolerance value that we can be 90% confident that 95% of the products produced will be above?

$$X_L = \bar{X} - K_1 s$$

$$1.500 - 1.890 * 0.002 = 0.0038$$

$$1.500 - 0.0038 = 1.4962$$

We can be 95% confident that 99% of the parts are above 1.4692.

Example: A process has an average of 1.500 and a sample standard deviation of 0.002. If 80 parts are produced, what is the tolerance value that we can be 90% confident that 95% of the products produced will be below?

$$X_U = \bar{X} + K_1 s$$

$$1.500 + 1.890 * 0.002 = 0.0038$$

$$1.500 + 0.0038 = 1.5038$$

We can be 95% confident that 99% of the parts are below 1.5038.

3.2 PREDICTION INTERVALS

Unlike confidence intervals, which look at population parameters such as the mean and standard deviation, *prediction intervals* are used to predict the possible value of a future observation. Prediction intervals assume an underlying normal distribution.

3.3 PREDICTION INTERVALS WHEN THE VARIANCE (σ^2) IS KNOWN

For measurements that are normally distributed with an unknown μ and a known variance σ^2, the two-sided prediction interval calculation for a future observation is

$$\bar{X} \pm Z_{\alpha/2} * \sigma * \sqrt{1 + \frac{1}{n}}$$

where

σ = Standard deviation

$\bar{X}$ = The sample mean

n = Sample size

$Z_{\alpha/2}$ = The normal distribution value for a given confidence level (see Appendix D, Selected Single-Sided Normal Distribution Probability Points)

Example: A process has an average of 1.500 for the last 50 parts produced and a historical standard deviation of 0.002. Find the 95% prediction interval for the next part produced:

$$\bar{X} \pm Z_{\alpha/2} * \sigma * \sqrt{1 + \frac{1}{n}}$$

$$1.500 \pm 1.960 * 0.002 * \sqrt{1 + \frac{1}{50}} = 0.004$$

$$1.500 \pm 0.004 = 1.496 \text{ to } 1.504$$

We can be 95% confident the next part will fall between 1.496 and 1.504.

The upper-tailed prediction interval is given as follows:

$$\bar{X} - Z_{\alpha} * \sigma * \sqrt{1 + \frac{1}{n}}$$

where

σ = Standard deviation

$\bar{X}$ = The sample mean

n = Sample size

Z_{α} = The normal distribution value for a given confidence level (see Appendix C, Selected Single-Sided Normal Distribution Probability Points)

Example: A process has an average of 1.500 for the last 50 parts produced and a historical standard deviation of 0.002. Find the 95% prediction interval that the next part produced will be above:

$$\bar{X} - Z_{\alpha} * \sigma * \sqrt{1 + \frac{1}{n}}$$

$$1.500 - 1.645 * 0.002 * \sqrt{1 + \frac{1}{50}} = 0.003$$

$$1.500 - 0.003 = 1.497$$

We can be 95% confident the next part will be above 1.497.

The lower-tailed prediction interval is given as follows:

$$\bar{X} + Z_\alpha * \sigma * \sqrt{1 + \frac{1}{n}}$$

where

σ = Standard deviation

$\bar{X}$ = The sample mean

n = Sample size

Z_α = The normal distribution value for a given confidence level (see Appendix C, Selected Single-Sided Normal Distribution Probability Points)

Example: A process has an average of 1.500 for the last 50 parts produced and a historical standard deviation of 0.002. Find the 95% prediction interval that the next part produced will be below:

$$\bar{X} + Z_\alpha * \sigma * \sqrt{1 + \frac{1}{n}}$$

$$1.500 + 1.645 * 0.002 * \sqrt{1 + \frac{1}{50}} = 0.003$$

$$1.500 + 0.003 = 1.503$$

We can be 95% confident the next part will be below 1.503.

3.4 PREDICTION INTERVALS WHEN THE VARIANCE (σ^2) IS UNKNOWN

For measurements that are normally distributed with an unknown μ and an unknown variance σ^2, the two-sided prediction interval calculation for a future observation is

$$\bar{X} \pm t_{\alpha/2,n-1} * s * \sqrt{\frac{n+1}{n}}$$

where

s = Sample standard deviation

$\bar{X}$ = The sample mean

n = Sample size

$t_{\alpha/2,n-1}$ = Percentage points of the Student's t distribution value for a given confidence level for $(n-1)$ degrees of freedom (see Appendix E, Percentage Points of the Student's t-Distribution)

Example: A process has an average of 1.500 for the last 30 parts produced and a sample standard deviation of 0.002. Find the 95% prediction interval for the next part produced:

$$\bar{X} \pm t_{\alpha/2,n-1} * s * \sqrt{\frac{n+1}{n}}$$

$$1.500 \pm 2.045 * 0.002 * \sqrt{\frac{60+1}{60}} = 0.004$$

$$1.500 \pm 0.004 = 1.496 \text{ to } 1.504$$

We can be 95% confident the next part will fall between 1.496 and 1.504. The upper-tailed prediction interval is given as follows:

$$\bar{X} - t_{\alpha,n-1} * \sigma * \sqrt{\frac{n+1}{n}}$$

where

s = Sample standard deviation

$\bar{X}$ = The sample mean

n = Sample size

$t_{\alpha,n-1}$ = Percentage points of the Student's t distribution value for a given confidence level for $(n-1)$ degrees of freedom (see Appendix E, Percentage Points of the Student's t-Distribution)

Example: A process has an average of 1.500 for the last 30 parts produced and a sample standard deviation of 0.002. Find the 90% prediction interval that the next part produced will be above:

$$\bar{X} - t_{\alpha,n-1} * \sigma * \sqrt{\frac{n+1}{n}}$$

$$1.500 - 1.699 * 0.002 * \sqrt{\frac{30+1}{30}} = 0.003$$

$$1.500 - .003 = 1.497$$

We can be 90% confident the next part will be above 1.497.

The lower-tailed prediction interval is given as follows:

$$\bar{X} + t_{\alpha,n-1} * \sigma * \sqrt{\frac{n+1}{n}}$$

where

σ = Sample deviation

$\bar{X}$ = The sample mean

n = Sample size

$t_{\alpha,n-1}$ = Percentage points of the Student's t distribution value for a given confidence level for $(n-1)$ degrees of freedom (see Appendix E, Percentage Points of the Student's t-Distribution)

Example: A process has an average of 1.500 for the last 30 parts produced and a historical standard deviation of 0.002. Find the 90% prediction interval that the next part produced will be below:

$$\bar{X} + t_{\alpha,n-1} * \sigma * \sqrt{\frac{n+1}{n}}$$

$$1.500 + 1.699 * 0.002 * \sqrt{\frac{30+1}{30}} = 0.003$$

$$1.500 + 0.003 = 1.503$$

We can be 90% confident the next part will be below 1.503.

4

Correlation and Regression Analysis

C _orrelation analysis_ is a method for determining the strength of a linear relationship of two variables. While regression analysis helps to assess the association and relationship between two variables, it is strongly recommended that a visual review of the plotted data points is conducted to look for the presence of a linear relationship prior to performing the calculations. If the plotted points generally do not form a straight line (that is, curves, or other nonlinear patterns) the calculations will be invalid and therefore not meaningful. These techniques are generally used to quantify the results of scatter plots.

4.1 CORRELATION ANALYSIS

Correlation analysis is the measure of strength or the degree of linear association between two variables. The correlation coefficient can vary from positive 1 (indicating a perfect positive relationship), through zero (indicating the absence of a relationship), to negative 1 (indicating a perfect negative relationship). As a rule of thumb, correlation coefficients between .00 and .30 are considered weak, those between .30 and .70 are moderate, and coefficients between .70 and 1.00 are considered high. However, this rule should always be qualified by a visual inspection of the plotted points. Figure 4.1 is a visual representation of the relative degrees of correlation.

Table 4.1 presents a basic review of the function and purpose of the X and Y variables.

The correlation coefficient can be calculated using the following formula:

$$r = \frac{n * \Sigma XY - \Sigma X * \Sigma Y}{\sqrt{n * \Sigma X^2 - (\Sigma x)^2} * \sqrt{n * \Sigma Y^2 - (\Sigma Y)^2}}$$

where

r = Correlation coefficient

X = The independent variable

Y = The dependent variable

n = The number of sample pairs

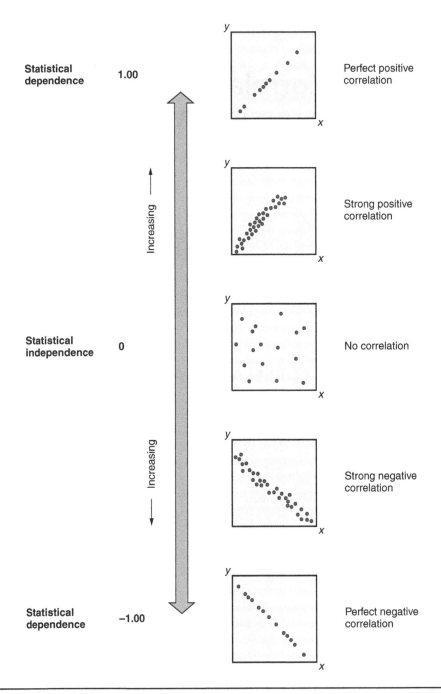

Figure 4.1 Relative degrees of correlation.

Table 4.1   Correlation axis table.

X-axis	Y-axis
Independent	Dependent
Predictor	Predicted
Carrier	Response
Input	Output

Table 4.2   Example calculation summary.

Time	Strength			
X	Y	XY	X^2	Y^2
1.0	41	41	1.0	1681
1.5	41	61.5	2.3	1681
2.0	41	82	4.0	1681
2.5	43	107.5	6.3	1849
3.0	43	129	9.0	1849
3.5	44	154	12.3	1936
4.0	50	200	16.0	2500
17.5	**303.0**	**775.0**	**50.8**	**13177.0**	Sum

Example: A study was done to compare the strength of a molded resin versus the cure time in the processing oven. The data points were recorded, as shown in Table 4.2.

$$r = \frac{n * \Sigma XY - \Sigma X * \Sigma Y}{\sqrt{n * \Sigma X^2 - (\Sigma x)^2} * \sqrt{n * \Sigma Y^2 - (\Sigma Y)^2}}$$

$$r = \frac{7 * 775 - 17.5 * 303}{\sqrt{7 * 50.8 - (17.5)^2} * \sqrt{7 * 13177 - (303)^2}} = 0.843$$

The correlation coefficient is 0.843, which indicates a strong positive relationship. The coefficient of determination r^2 is simply the correlation coefficient squared. In this case the coefficient of determination is 0.711, indicating that 71.1% of the variation of strength (*Y*) is explained by the variation in time (*X*).

　　To further test the significance of the calculated correlation coefficient, the calculated value can be compared to a correlation coefficient critical value (see Appendix I, Critical Values of the Correlation Coefficient). If the absolute value of *r* is greater than the critical table value, then it can be stated that you are confident that there is a statistically significant relationship between the *X* and *Y* variables.

The correlation coefficient is 0.843, and we wish to be 95% confident of the relationship between the variables X and Y. Looking at Appendix I, Critical Values of the Correlation Coefficient, 5 degrees of freedom $(7 - 2)$ and an α of 0.05 yields a value of 0.754. We can be 95% sure there is a statistically significant relationship between the X and Y variables.

Please note that a graphic plot of the data points would confirm the calculations.

4.2 REGRESSION ANALYSIS

Regression analyses assess the association between two variables and define their relationship. The general equation for a line is given by:

$$Y = aX + b$$

where

$$a = \frac{n * \Sigma XY - \Sigma X * \Sigma Y}{n * \Sigma X^2 - (\Sigma X)^2}$$

and

$$b = \frac{\Sigma Y - a * \Sigma X}{n}$$

where

X = The independent variable

Y = The dependent variable

n = The number of sample pairs

a = The slope of the regression line

b = The intercept point on the Y axis

Example: Using the information from Table 4.2, calculate the regression equation:

$$a = \frac{n * \Sigma XY - \Sigma X * \Sigma Y}{n * \Sigma X^2 - (\Sigma X)^2} = \frac{7 * 775 - 17.5 * 303}{7 * 50.8 - 17.5^2} = 2.48$$

and

$$b = \frac{\Sigma Y - a * \Sigma X}{n} = \frac{303 - 2.48 * 17.5}{7} = 37.086$$

Putting the calculated values into the standard form will yield the following regression line:

$$Y = 2.48X + 37.086$$

Please note that a graphic plot of the data points would confirm the calculations. It must also be noted that predictions outside of the data values used to calculate the regression line must be used with caution.

4.3 NORMAL PROBABILITY PLOTS

Normal probability plots can be constructed to look for linearity when using one variable. Normal probability plots give a visual way to determine if a distribution is approximately normal.

If the distribution is close to normal, the plotted points will lie close to a line. Systematic deviations from a line indicate a nonnormal distribution:

- *Right skew.* If the plotted points appear to bend up and to the left of the normal line, this indicates a long tail to the right (Figure 4.2).

- *Left skew.* If the plotted points bend down and to the right of the normal line, this indicates a long tail to the left (Figure 4.3).

- *Short tails.* An S-shaped curve indicates shorter-than-normal tails, that is, less variance than expected (Figure 4.4).

Figure 4.2   Right-skewed distribution.

Figure 4.3   Left-skewed distribution.

Figure 4.4 Short-tailed distribution.

Figure 4.5 Long-tailed distribution.

- *Long tails.* A curve that starts below the normal line, bends to follow it, and ends above it indicates long tails. That is, you are seeing more variance than you would expect in a normal distribution (Figure 4.5).

Normal probability plots are constructed by doing the following:

1. The data are arranged from smallest to largest.

2. The percentile of each data value is determined.

3. From these percentiles, normal calculations are done to determine their corresponding z-scores.

4. Each z-score is plotted against its corresponding data value.

To calculate the percentile, use the following formula:

$$p = \frac{(i - 0.5)}{n}$$

where

i = The individual value

n = The number of data points

Each p value must be converted to its corresponding z-score using Appendix A Normal Distribution Probability Points – Area below Z for p values ≥ 0.500 and Appendix B Normal Distribution Probability Points – Area above Z for p values ≤ 0.500. For example p 0.07 using Appendix B, $Z = -1.47$, p 0.93 using Appendix A, $z = 1.47$. In these cases

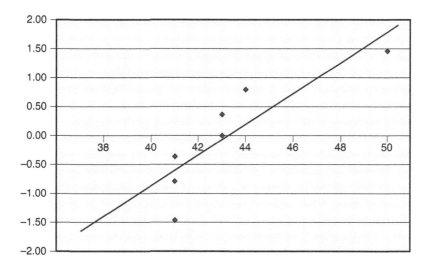

Figure 4.6 Normal probability plot for strength.

Table 4.3 Calculation summary.

Strength	i	p	z
41	1	0.07	−1.47
41	2	0.21	−0.79
41	3	0.36	−0.37
43	4	0.50	0.00
43	5	0.64	0.37
44	6	0.79	0.79
50	7	0.93	1.47

there was not an exact z value which corresponds to the p value. So one must chose a z-value closest to the p value. For any p value below 0.50 the z will be negative and for any p value above 0.50 the z value will be positive.

A graphical review of the normal probability plot for strength (Figure 4.6) does not yield a normal distribution for the data points analyzed as they do not generally fall on the line of best fit.

5

Outliers

O*utliers* are data points that do not appear to "belong" to a given set of data. Outliers can be caused by recording and measurement errors, incorrect distribution assumption, unknown data structure, or novel phenomenon (Iglewicz 1993). Additionally, outliers can be created by a shift in the mean or a change in the variation. There are several tools both visual (histogram, box plot, and so on) and analytical that can help to determine whether a suspect point is truly an outlier and should be removed from consideration when analyzing a data set.

Once an observation is identified (by means of a graphical or visual inspection) as a potential outlier, root cause analysis should begin to determine whether an assignable cause can be found for the spurious result (Walfish 2006).

5.1 OUTLIER DETECTION BASED ON THE STANDARD DEVIATION FOR A NORMAL DISTRIBUTION

One of the easiest methods for detecting an outlier is to use the standard deviation method. If the data are normally distributed, a single value may be considered an outlier if it falls outside of ± 3σ (approximately 99.8% of normally distributed data fall within this range):

$$z = \frac{|x_i - \mu|}{\sigma}$$

where

μ = Process average

x_i = Potential outlier

σ = Standard deviation

If $z_{Calc} > 3$, then the point can be considered an outlier.

Assuming data from a process are normally distributed, with a historical μ of 8.5 and a σ of 2.3 (Figure 5.1), determine if the following two points can be considered outliers: 1 and 19 (±3σ ≈ 99.8%).

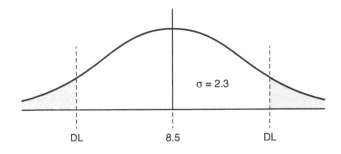

Figure 5.1   Normal distribution.

Example: Testing to see if the value of 1 is an outlier:

$$z = \frac{|x_i - \mu|}{\sigma} = \frac{|1 - 8.5|}{2.3} = 3.26$$

Since 3.26 is greater than 3, we can conclude that the value of 1 is an outlier, as a value of 1 would be expected to occur only 0.1% of the time.

Example: Testing to see if the value of 19 is an outlier:

$$z = \frac{|x_i - \mu|}{\sigma} = \frac{|19 - 8.5|}{2.3} = 4.56$$

Since 4.56 is greater than 3, we can conclude that the value of 19 is an outlier, as a value of 19 would be expected to occur only 0.1% of the time.

5.2 DISCORDANCE OUTLIER TEST

The *discordance test* for outliers is similar to the standard deviation method with the exception that a *D* statistic is calculated and compared to a table value. This test assumes a normal distribution:

$$D = \frac{|\mu - x_i|}{\sigma}$$

where

D = Discordance value

μ = Process average

x_i = Potential outlier

σ = Standard deviation

$D_{(\alpha,n)}$ = Critical value for a given confidence level and sample size (n)
      (see Appendix L, Critical Values for the Discordance Outlier Test)

Example: Assuming data from a process are normally distributed, with a historical μ of 8.5, σ of 2.3, and $n = 37$, determine if the following two points can be considered outliers: 1 and 19 (at the 0.05 α level).
      Testing to see if 1 is an outlier:

$$D = \frac{|\mu - x_i|}{\sigma} = \frac{8.5 - 1}{2.3} = 3.26 \quad (D_{37\alpha=0.05} \text{ table value is 2.835})$$

Since D_{Calc} 3.26 > D_{Crit} 2.835, we are 95% sure the point an outlier.
      Testing to see if 19 is an outlier:

$$D = \frac{|\mu - x_i|}{\sigma} = \frac{8.5 - 19}{2.3} = 4.56 \quad (D_{37\alpha=0.05} \text{ table value is 2.835})$$

Since D_{Calc} 4.56 > D_{Crit} 2.835, we are 95% sure the point an outlier.

5.3 OUTLIER DETECTION BASED ON THE STANDARD DEVIATION FOR AN UNKNOWN DISTRIBUTION

Chebyshev's inequality theorem can be used to determine if a point is an outlier when the underlying distribution is unknown. This method can also be used with a normal distribution; however, this method is not as sensitive as the standard deviation method shown in Section 5.1. Chebyshev's theorem states that the portion of any set of data within k standard deviations of the mean is always at least $1 - (1/k^2)$, when k is > 1. Table 5.1 is a listing of k values and corresponding percentages.

Table 5.1 Selected k values.

k	%	k	%
1.25	36%	3.00	89%
1.50	56%	3.25	91%
1.75	67%	3.50	92%
2.00	75%	3.75	93%
2.25	80%	4.00	94%
2.50	84%	5.00	96%
2.75	87%	6.00	97%

$$C = \frac{|\mu - x_i|}{k\sigma}$$

where

C = Chebyshev value

μ = Process average

k = Value from Table 5.1

x_i = Potential outlier

σ = Standard deviation

If $C_{\text{Calc}} > 1$, then the point can be considered an outlier.

Example: Assuming data are from an unknown distribution from a process with a historical μ of 8.5, σ of 2.3, and a k value of 3 (89%) (Figure 5.2), determine if the following two points can be considered outliers: 1 and 19.

Testing to see if the value of 1 is an outlier:

$$C = \frac{|\mu - x_i|}{k\sigma} = \frac{|8.5 - 1|}{3 * 2.3} = 1.09$$

Since 1.09 is greater than 1, we can conclude that the value of 1 is an outlier.

Testing to see if the value of 19 is an outlier:

$$C = \frac{|\mu - x_i|}{k\sigma} = \frac{|8.5 - 19|}{3 * 2.3} = 1.52$$

Since 1.52 is greater than 1, we can conclude that the value of 19 is an outlier.

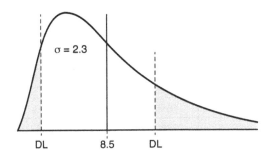

Figure 5.2   Unknown distribution.

5.4 OUTLIER DETECTION BASED ON THE INTERQUARTILE RANGE

Another method for detecting an outlier when the underlying distribution is unknown and assumed to be nonnormal is the *interquartile range method*. If a suspected point is at least 1.5 interquartile ranges below the first quartile ($Q1$) or at least 1.5 interquartile ranges above the third quartile ($Q3$), the point may be considered an outlier:

$$\text{LPO} = Q1 - 1.5(Q3 - Q1) \text{ and UPO} = Q3 + 1.5(Q3 - Q1)$$

where

LPO = Lower potential outlier

UPO = Upper potential outlier

$Q1$ = Median value in the first half of the data set

$Q3$ = Median value in the second half of the data set

Example: Given the following data set (please note that the data points should be ordered from smallest value to largest value), calculate the lower and upper outlier limits and determine which points, if any, may be considered an outlier:

$$1, 5, \mathbf{6}, 7, 8, 9, 10, \mathbf{11}, 12, 19$$

$X_{0.25} = 6$ (median value of the first half of the data set)

$X_{0.75} = 11$ (median value of the second half of the data set)

$$\text{LPO} = Q1 - 1.5(Q3 - Q1) = 6 - 1.5(11 - 6) = -1.5$$

and

$$\text{UPO} = Q3 + 1.5(Q3 - Q1) = 11 + 1.5(11 - 6) = 18.5$$

We can reasonably conclude that any point ≤ -1.5 and ≥ 18.5 may be considered an outlier. From the data set, the data point 19 is considered to be an outlier.

5.5 DIXON'S Q TEST

Dixon's Q test is a method that compares the gap (data point in question to the next closest value) divided by the range to determine if the suspect point is an outlier. It is recommended to use this method only once on a particular set of data. Once the Q value has been calculated, it must be compared to the critical value from Table 5.2. If Q calculated $> Q$ critical, the data point may be considered an outlier with the chosen confidence:

Table 5.2 Selected critical Q values.

n	Q critical 90%	Q critical 95%	Q critical 99%
3	0.941	0.970	0.994
4	0.765	0.829	0.926
5	0.642	0.710	0.821
6	0.560	0.625	0.740
7	0.507	0.568	0.680
8	0.468	0.526	0.634
9	0.437	0.493	0.598
10	0.412	0.466	0.568

$$Q_{Calc} = \frac{\text{Gap}}{\text{Range}}$$

Example: Given the following data set (please note that the data points should be ordered from smallest value to largest value), we suspect that the value of 17 may be an outlier. Calculate the Q statistic to determine if the point may be considered an outlier at the 95% confidence level:

$$1, 3, 6, 7, 8, 9, 10, 11, 12, 23$$

$$Q_{Calc} = \frac{\text{Gap}}{\text{Range}} = \frac{23-12}{23-1} = 0.500 \ (Q_{Crit\ \alpha=0.05} \text{ table value is } 0.466)$$

Since Q_{Calc} 0.500 > Q_{Crit} 0.466, we are 95% sure the point is an outlier.

5.6 DEAN AND DIXON OUTLIER TEST

The *Dean and Dixon outlier test* is a valid method for detecting outliers when the data are normally distributed. To use this test, the data must be ordered from smallest value to largest value. The formulas are dependent on the sample size:

N	Smallest value test	Largest value test
3 to 7	$r_{10} = \dfrac{X_2 - X_1}{X_n - X_1}$	$r_{10} = \dfrac{X_n - X_{n-1}}{X_n - X_1}$
8 to 10	$r_{11} = \dfrac{X_2 - X_1}{X_{n-1} - X_1}$	$r_{11} = \dfrac{X_n - X_{n-1}}{X_n - X_2}$

$$11 \text{ to } 13 \qquad r_{21} = \frac{X_3 - X_1}{X_{n-1} - X_1} \qquad\qquad r_{21} = \frac{X_n - X_{n-2}}{X_n - X_2}$$

$$14 \text{ to } 30 \qquad r_{22} = \frac{X_3 - X_1}{X_{n-2} - X_1} \qquad\qquad r_{22} = \frac{X_n - X_{n-2}}{X_n - X_3}$$

where

n = Number of data points

(See Appendix J, Critical Values of the Dean and Dixon Outlier Test.)

Example: Given the following data points, verify whether the smallest and largest values are outliers at 95% confidence:

$$1, 3, 6, 7, 8, 9, 10, 11, 12, 23$$

Since there are 10 data points ($n = 10$) we must use the second set of equations for 8 to 10 to detect the presence of an outlier.

To detect whether the smallest value is an outlier:

$$r_{11} = \frac{X_2 - X_1}{X_{n-1} - X_1} = \frac{3-1}{12-1} = 0.182 \ (r_{11\,\alpha=0.05} \text{ table value is } 0.477)$$

Since r_{11} calc is < r_{11} crit, we can conclude that the smallest value is not an outlier at the 95% confidence level.

To detect whether the largest value is an outlier:

$$r_{11} = \frac{X_n - X_{n-1}}{X_n - X_2} = \frac{23-12}{23-3} = 0.550 \ (r_{11\,\alpha=0.05} \text{ table value is } 0.477)$$

Since r_{11} calc is > r_{11} crit we can conclude the largest value is an outlier at the 95% confidence level.

5.7 GRUBBS OUTLIER TEST

The *Grubbs outlier test* computes the outlying data point using the average and standard deviation and then compares the calculated value against a critical table value. When a data point is deemed an outlier and removed from the data set, and an additional outlier is suspected, the average and standard deviation must be recalculated.

We calculate the g statistic using

$$g_{\text{Min}} = \frac{\bar{X} - x_{\text{Min}}}{s} \quad \text{or} \quad g_{\text{Max}} = \frac{x_{\text{Max}} - \bar{X}}{s}$$

where

$\bar{X}$ = Average

s = Standard deviation

x_{Min} = Suspected minimum value

x_{Max} = Suspected maximum value

(See Appendix K, Critical Values for the Grubbs Outlier Test.)

Example: Given the following data points, verify whether the smallest and largest values are outliers at 95% confidence:

$$1, 3, 6, 7, 8, 9, 10, 11, 12, 23 \ (\bar{X} = 9, s = 6, \text{ and } n = 10)$$

$$g_{Min} = \frac{\bar{X} - x_{Min}}{s} = \frac{9 - 1}{6} = 1.333 \ (g_{Crit \, \alpha = 0.05} \text{ table value is } 2.176)$$

Since g_{Min} calc is $< g_{Crit}$, we can conclude that the smallest value is not an outlier at the 95% confidence level.

$$g_{Max} = \frac{x_{Max} - \bar{X}}{s} = \frac{23 - 9}{6} = 2.333 \ (g_{Crit \, \alpha = 0.05} \text{ table value is } 2.176)$$

Since g_{Max} calc is $> g_{Crit}$, we can conclude that the largest value is an outlier at the 95% confidence level.

5.8 WALSH'S OUTLIER TEST

Walsh's outlier test is a nonparametric test that can be used to detect multiple outliers when the data are not normally distributed. This test requires a minimum sample size of 60. When the sample size (n) is between 60 and 219, the α level of significance is 0.10. When the sample size (n) is 220 or larger, the α level of significance is 0.05. To begin the process, the data must be ordered from smallest to largest. This test requires several calculations and is somewhat cumbersome:

$$c = \sqrt{2n}$$

$$k = r + c$$

$$b^2 = \frac{1}{\alpha}$$

$$a = \frac{1 + b\sqrt{(c - b^2)/(c-1)}}{c - b^2 - 1}$$

rounds up to next whole integer.

Where r smallest points are outliers:

$$X_{(r)} - (1+a)X_{(r+1)} + aX_{(k)} < 0$$

Where r largest points are outliers:

$$X_{(n+1-r)} - (1+a)X_{(n-r)} + aX_{(n+1-k)} > 0$$

Example: 250 parts were measured. It is suspected that the two smallest and the two largest parts are potential outliers. Because $n > 220$, it is appropriate to use an α significance level of 0.05:

$$c = \sqrt{2n} = \sqrt{2 * 250} = 22.361$$

$$k = r + c = 2 + 22.361 = 24.361$$

$$b^2 = \frac{1}{\alpha} = \frac{1}{.05} = 20$$

$$a = \frac{1 + b\sqrt{(c - b^2)/(c-1)}}{c - b^2 - 1} = \frac{1 + 4.472\sqrt{(22.361 - 20)/(22.361 - 1)}}{22.361 - 20 - 1} = 1.829$$

$$a = 1.829 \text{ rounded up} = 2.0$$

Where r smallest points are outliers:

$$X_{(r)} - (1+a)X_{(r+1)} + aX_{(k)} < 0 = 7_2 - (1 + 2)10_3 + 2 * 42_{23} < 0$$

$$25 < 0$$

Since 25 is not less than 0, we are 95% confident the smallest two values are outliers.

where

$X_2 = 7$

$X_3 = 10$ (values from the data set ordered)

$X_{23} = 42$

$$X_{(r)} - (1+a)X_{(r+1)} + aX_{(k)} < 0$$

Where r largest points are outliers:

$$X_{(n+1-r)} - (1+a)X_{(n-r)} + aX_{(n+1-k)} > 0 = 214_{499} - (1+2)210_{498} + 275_{226} > 0$$

$$-141 > 0$$

where

$X_{499} = 214$

$X_{498} = 210$ (values from the data set ordered)

$X_{226} = 75$

Since -141 is less than 0, we are 95% confident the three largest values are not outliers.

5.9 HAMPEL'S METHOD FOR OUTLIER DETECTION

Because the mean and standard deviation are adversely influenced by the presence of outliers, Hampel's method is presented (Hampel 1971; Hampel 1974). For example, the mean will be offset toward the outlier, and the standard deviation will be inflated, leading to errant statistical decisions. Hampel's method is somewhat resistant to these issues as the calculations use the median and *median absolute deviation* (MAD) to detect the presence of outliers.

To determine which, if any, values are outliers, calculate the median value ($\tilde{x}$) (see Section 1.1, Estimates of Central Tendency for Variables Data), calculate the $\text{MAD} = |x_i - \tilde{x}|$, and finally calculate the decision value (median MAD * 5.2). Compare the calculated MAD values to the decision value. Any MAD that exceeds the decision value can be considered an outlier:

$$\text{MAD} = |x_i - \tilde{x}|$$

Decision value (median MAD * 5.2)

Example: The following data were recorded from a manufacturing process with an unknown distribution: 1, 5, 6, 7, 8, 9, 10, 11, 12, and 19. Determine whether any of these values can be considered outliers:

$$\tilde{x} = \frac{8+9}{2} = 8.5$$

	MAD
Value	$\lvert x_i - \tilde{x}\rvert$
1	$\lvert 1 - 8.5 \rvert = 7.5$
5	$\lvert 5 - 8.5 \rvert = 3.5$
6	$\lvert 6 - 8.5 \rvert = 2.5$
7	$\lvert 7 - 8.5 \rvert = 1.5$
8	$\lvert 8 - 8.5 \rvert = 0.5$
9	$\lvert 9 - 8.5 \rvert = 0.5$
10	$\lvert 10 - 8.5 \rvert = 1.5$
11	$\lvert 11 - 8.5 \rvert = 2.5$
12	$\lvert 12 - 8.5 \rvert = 3.5$
19	$\lvert 19 - 8.5 \rvert = 10.5$

Value	**MAD**
8	0.5
9	0.5
7	1.5
10	1.5
6	2.5
11	2.5
5	3.5
12	3.5
1	7.5
19	10.5

$$\text{median MAD} = \frac{2.5 + 2.5}{2} = 2.5$$

Decision value (median MAD $* 5.2$) $= 2.5 * 5.2 = 13$

Since none of the MAD values exceed the decision value of 13, none of the values are considered outliers.

6

Hypothesis Testing

6.1 TYPE I AND TYPE II ERRORS

When testing hypotheses, we usually work with small samples from large populations. Because of the uncertainties of dealing with sample statistics, decision errors are possible. There are two types of errors: type I errors and type II errors.

A *type I error* occurs when the null hypothesis is rejected when it is, in fact, true. In most tests of hypotheses, the risk of committing a type I error (known as the *alpha* [α] *risk*) is stated in advance along with the sample size to be used (see Table 6.1).

A *type II error* occurs when the null hypothesis is accepted when it is false. If the means of two processes are, in fact, different, and we accept the hypothesis that they are equal, we have committed a type II error. The risk associated with type II errors is referred to as the *beta (β) risk*. This risk is different for every alternative hypothesis and is usually unknown except for specific values of the alternative hypothesis (see Table 6.1).

We cover these risks in more detail in Chapter 7, Sample Size Determination for Tests of Hypotheses.

6.2 ALPHA (α) AND BETA (β) RISKS

There is an associated risk with each of the errors just described.

The risk of rejecting the null hypothesis when it is true is the *alpha (α) risk*. The experimenter usually determines the risk before running the test. Common levels of alpha (α) risks are 0.10, 0.05, and 0.01.

Table 6.1 Hypothesis truth table.

Decision on null hypothesis	Null hypothesis is true	Null hypothesis is false
Accept (Fail to reject)	Correct decision Probability = $1 - \alpha$	Type II error Probability = β
Reject	Type I error Probability = α	Correct decision Probability = $1 - \beta$

An alpha (α) risk of 0.10 means that, in the long run, the null hypothesis will be rejected when it is true (type I error) 10 times out of a hundred; an alpha (α) risk of 0.05 means that, in the long run, the null hypothesis will be rejected when it is true (type I error) five times out of a hundred; and an alpha (α) risk of 0.01 means that, in the long run, the null hypothesis will be rejected when it is true (type I error) one time out of a hundred, and so on.

Type I and type II errors can not occur at the same time. Type I error can happen only if H_0 is true, and type II error can happen only when H_0 is false. If the probability of type I error is high, then the probability of type II error is low.

The statistical power of a test is also another consideration when conducting tests of hypotheses. The power of a test is $1 - \beta$, and is expressed as a probability ranging from 0 to 1. The higher the power, the more sensitive the test is in detecting differences where differences exist between the hypothesized parameter and its true value. Increasing the sample size, reducing the variance, increasing the effect size index, or decreasing the alpha level (α) will increase the statistical power.

6.3 THE EFFECT SIZE INDEX

The *effect size* is the degree of departure of the alternative hypothesis from the null hypothesis. It is indexed to obtain a "pure number," *d*, free of the original measurement unit, by dividing the difference of the observed results by their standard deviation:

$$d = \left| \frac{m_1 - m_2}{\sigma} \right|$$

where

m_1 = The mean of group 1

m_2 = The mean of group 2

σ = Standard deviation of either one of the means

Example: A change had been made to the molding process of a plastic part. We wish to analyze the effect of the change. Using the original settings, the mean is 25 with a σ of 5; with the modified settings, the mean is 20 with a σ of 3. Calculate the effect size:

$$d = \left| \frac{m_1 - m_2}{\sigma} \right| = \left| \frac{25 - 20}{5} \right| = 1$$

Referring to the following table, an effect size of 1 is considered to be a large effect.

$0 < d < 0.2$	Small effect
$0.2 < d < 0.8$	Medium effect
$d > 0.8$	Large effect

When we wish to pool the standard deviations, the following formula may be used:

$$d = \left| \frac{m_1 - m_2}{\sqrt{(\sigma_{m_1}^2 + \sigma_{m_2}^2)/2}} \right|$$

where

m_1 = The mean of group 1

m_2 = The mean of group 2

$\sigma_{m_1}^2$ = Standard deviation of group 1

$\sigma_{m_2}^2$ = Standard deviation of group 2

Example: An additional change had been made to the molding process of a plastic part. We wish to analyze the effect of the change. Using the original settings, the mean is 25 with a σ of 5; with the modified settings, the mean is 24 with a σ of 3. Calculate the effect size:

$$d = \left| \frac{m_1 - m_2}{\sqrt{(\sigma_{m_1}^2 + \sigma_{m_2}^2)/2}} \right| = \left| \frac{25 - 24}{\sqrt{(25 + 9)/2}} \right| = \frac{1}{4.123} = 0.243$$

Referring to the above table, an effect size of 0.243 is considered to be a medium effect.

6.4 APPORTIONMENT OF RISK IN HYPOTHESIS TESTING

Tests of hypotheses and the apportionment of the risk usually fall into two categories: one-tail tests, where all of the risk is in a single tail of the distribution, and two-tail tests, where the risk is evenly split between the two tails of the distribution. When we conduct tests of hypotheses, we must calculate a value and compare it to a critical value from a reference table. It is possible we can fail to reject the double-tailed test, the upper-tailed test, and the lower-tailed test null hypotheses.

6.5 THE HYPOTHESIS TEST FOR A ONE-TAIL (UPPER-TAILED) TEST

The hypothesis for a one-tail (upper-tailed) test might be stated as

$$\begin{array}{cc} H_0: \ _1 \leq \ _2 & H_0: \sigma_1^2 \leq \sigma_2^2 \\ & \text{or} \\ H_A: \ _1 > \ _2 & H_A: \sigma_1^2 > \sigma_2^2 \end{array}$$

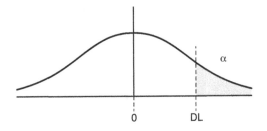

Figure 6.1   Representation of a one-tail (upper-tailed) test.

The null hypothesis (H_0) is that $\mu_1 \le \mu_2$ or $H_0 : \sigma_1^2 \le \sigma_2^2$, while the alternative hypothesis (H_A) is that $\mu_1 > \mu_2$ or $H_A : \sigma_1^2 > \sigma_2^2$.

We reject the null hypothesis (H_0) if $Z_{\text{Calculated}} > Z_{\text{Critical}}$.

The hypothesis for a one-tail (upper-tailed) test, as depicted in Figure 6.1, is used when one suspects that the alternative hypothesis (H_A) is greater than the null hypothesis (H_0).

6.6 THE HYPOTHESIS TEST FOR A ONE-TAIL (LOWER-TAILED) TEST

The hypothesis for a one-tail (lower-tailed) test might be stated as

$$H_0 : \mu_1 \ge \mu_2 \qquad H_0 : \sigma_1^2 \ge \sigma_2^2$$
$$\text{or}$$
$$H_A : \mu_1 < \mu_2 \qquad H_A : \sigma_1^2 < \sigma_2^2$$

The null hypothesis (H_0) is that $\mu_1 \ge \mu_2$ or $H_0 : \sigma_1^2 \ge \sigma_2^2$, while the alternative hypothesis (H_A) is that $\mu_1 < \mu_2$ or $H_A : \sigma_1^2 < \sigma_2^2$.

We reject the null hypothesis (H_0) if $Z_{\text{Calculated}} < Z_{\text{Critical}}$.

The hypothesis for a one-tail (lower-tailed) test, as depicted in Figure 6.2, is used when one suspects that the alternative hypothesis (H_A) is less than the null hypothesis (H_0).

6.7 THE HYPOTHESIS TEST FOR A TWO-TAIL TEST

The hypothesis for a two-tail test might be stated as

$$H_0 : \mu_1 = \mu_2 \qquad H_0 : \sigma_1^2 = \sigma_2^2$$
$$\text{or}$$
$$H_A : \mu_1 \ne \mu_2 \qquad H_A : \sigma_1^2 \ne \sigma_2^2$$

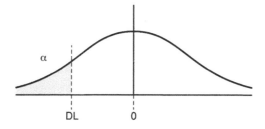

Figure 6.2 Representation of a one-tail (lower-tailed) test.

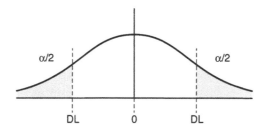

Figure 6.3 Representation of a two-tail test.

The null hypothesis (H_0) is that $\mu_1 = \mu_2$ or $H_A : \sigma_1^2 \neq \sigma_2^2$, while the alternative hypothesis (H_A) is that $\sigma_1 \neq \sigma_2$ or $H_A : \sigma_1^2 \neq \sigma_2^2$.

We reject the null hypothesis (H_0) if $\left| Z_{\text{Calculated}} \right| > \pm Z_{\text{Critical}}$.

The hypothesis for a two-tail (double-tailed) test, as depicted in Figure 6.3, is used when one suspects that the alternative hypothesis (H_A) is not equal to the null hypothesis (H_0).

6.8 THE HYPOTHESIS TEST CONCLUSION STATEMENTS

Once a hypothesis test has been conducted, it is necessary to draw a conclusion and formulate a summary statement of the findings.

Example 1: *(Fail to reject the null hypothesis).* Since Z calculated is less than Z critical at the 0.05 α (alpha) level, there is insufficient evidence to reject the null hypothesis that H_0: $\mu_1 = \mu_2$.

Example 2: *(Reject the null hypothesis).* Since Z calculated is greater than Z critical at the 0.05 α (alpha) level, there is sufficient evidence to reject the null hypothesis H_0: $\sigma_1 = \sigma_2$ in favor of the alternative hypothesis that $H_A : \sigma_1^2 \neq \sigma_2^2$.

7

Sample Size Determination for Tests of Hypotheses

Sometimes, when testing for a difference between a population mean versus a hypothesized mean, or when testing two population means, controlling only the α risk is not sufficient. For example, if the processing cost of a part is very high, and a small but real reduction in processing time can save considerable costs, we want to ensure that our test reveals this reduction.

If the processing time averages 28 hours with a standard deviation of 1.5 hours, a 1.5-hour reduction in processing time (a difference of one standard deviation) with no change in variation would result in a 5.4% savings in time, as well as other cost savings. If this were the case, it would be beneficial to set a low risk, called *beta* (β), of not detecting a change as small as one standard deviation.

In these situations, we must determine in advance the α and β risks and the difference to detect in units of standard deviation. Then, using methods from Natrella (1966), we can determine the sample size required to ensure that these risks are maintained.

Relatively few equations will be expanded or modified to determine the required sample size. The difference in means that are tested in the hypothesis at a given α and β risk will be in the number of standard deviations.

7.1 SAMPLE SIZE REQUIRED TO TEST AN OBSERVED MEAN VERSUS A HYPOTHESIZED MEAN WHEN STANDARD DEVIATION (σ) IS KNOWN

When performing the one-tail or two-tail test, we must determine the difference in standard deviation units (d):

$$d = \left| \frac{\mu_1 - \mu_2}{\sigma} \right|$$

where

μ_1 = The mean value that we want to detect at the given risk level

μ_2 = The hypothesized mean

σ = Standard deviation

The required sample size for a two-tail test is calculated by

$$n = \frac{(Z_{1-\alpha/2} + Z_{1-\beta})^2}{d^2}$$

where

$Z_{1-\alpha/2}$ = The normal distribution value for a given confidence level (see Appendix D, Selected Double-Sided Normal Distribution Probability Points)

$Z_{1-\beta}$ = The normal distribution value for a given confidence level (see Appendix C, Selected Single-Sided Normal Distribution Probability Points)

Example: The current process has an average of 1.500 and a standard deviation of 0.06. A new method of processing is proposed, but a change in the process average of .024 in either direction would be detrimental, and we would like to detect a shift of this magnitude with 95% confidence (β = .05 and α =.05):

$$d = \left| \frac{\mu_1 - \mu_2}{\sigma} \right| = \left| \frac{1.524 - 1.500}{0.06} \right| \text{ or } \left| \frac{1.476 - 1.500}{0.06} \right| = 0.40$$

$$n = \frac{(Z_{1-\alpha/2} + Z_{1-\beta})^2}{d^2} = \frac{(1.960 + 1.645)^2}{(0.40)^2} = 81.23$$

rounded up to the next integer is 82.

The required sample size for a one-tail test is calculated by

$$n = \frac{(Z_{1-\alpha} + Z_{1-\beta})^2}{d^2}$$

where

$Z_{1-\alpha}$ = The normal distribution value for a given confidence level (see Appendix C, Selected Single-Sided Normal Distribution Probability Points)

$Z_{1-\beta}$ = The normal distribution value for a given confidence level (see Appendix C, Selected Single-Sided Normal Distribution Probability Points)

Example: The current process has an average of 1.500 and a standard deviation of .06. A new method of processing is proposed, but a decrease in the process average of 0.024 would be detrimental, and we would like to detect a shift of this magnitude with 95% confidence (β = 0.05 and α = 0.05):

$$d = \left| \frac{\mu_1 - \mu_2}{\sigma} \right| = \left| \frac{1.476 - 1.500}{0.06} \right| = 0.40$$

$$n = \frac{(Z_{1-\alpha} + Z_{1-\beta})^2}{d^2} = \frac{(1.645 + 1.645)^2}{(0.40)^2} = 67.65$$

rounded up to the next integer is 68.

7.2 SAMPLE SIZE REQUIRED TO TEST AN OBSERVED MEAN VERSUS A HYPOTHESIZED MEAN WHEN THE STANDARD DEVIATION (σ) IS ESTIMATED FROM OBSERVED VALUES

When estimating the standard deviation from sample data, the *t*-test is the appropriate test of hypothesis. The equations provided in Section 7.1 are appropriate, except for the addition of a constant for the α risks of .01 and .05. These constants, which come from Natrella (1966), are shown in Table 7.1.

7.3 SAMPLE SIZE REQUIRED TO TEST FOR DIFFERENCES IN TWO OBSERVED MEANS WHEN STANDARD DEVIATION (σ) FOR EACH POPULATION IS KNOWN

When performing the one-tail or two-tail test, we must determine *d*:

$$d = \frac{|\mu_1 - \mu_2|}{\sqrt{\sigma_1^2 + \sigma_2^2}}$$

Table 7.1   Hypothesis constant table for two-tail and one-tail tests.

Two-tail test	Constant
If $\alpha = 0.01$	Add 4 to the calculated sample size
If $\alpha = 0.05$	Add 2 to the calculated sample size
If $\alpha \geq 0.10$	None required
One-tail test	**Constant**
If $\alpha = 0.01$	Add 3 to the calculated sample size
If $\alpha = 0.05$	Add 4 to the calculated sample size
If $\alpha \geq 0.10$	None required

where

μ_1 = The mean value of population 1

μ_2 = The mean value of population 2

σ_1 = Standard deviation of population 1

σ_2 = Standard deviation of population 2

The required sample size for a two-tail test is calculated by

$$n = \frac{(Z_{1-\alpha/2} + Z_{1-\beta})^2}{d^2} + 2$$

where

$Z_{1-\alpha/2}$ = The normal distribution value for a given confidence level (see Appendix D, Selected Double-Sided Normal Distribution Probability Points)

$Z_{1-\beta}$ = The normal distribution value for a given confidence level (see Appendix C, Selected Single-Sided Normal Distribution Probability Points)

Example: We are using two processes to manufacture a product. As the product streams are mixed at a later point, we determine that a difference in the two averages of 0.024 or greater is unacceptable. If the standard deviation of process one is 0.0714, and the standard deviation of process two is .06, determine the required (equal for both process streams) sample sizes to detect a difference as low as 0.024 with $\alpha = 0.05$ and $\beta = 0.10$:

$$d = \frac{|\mu_1 - \mu_2|}{\sqrt{\sigma_1^2 + \sigma_2^2}} = \frac{0.024}{\sqrt{0.0714^2 + 0.06^2}} = 0.257$$

$$n = \frac{(Z_{1-\alpha/2} + Z_{1-\beta})^2}{d^2} + 2 = \frac{(1.960 + 1.282)^2}{0.257^2} + 2 = 161.1$$

rounded up to the next integer is 162.

7.4 SAMPLE SIZE REQUIRED TO TEST FOR DIFFERENCES IN TWO OBSERVED MEANS WHEN THE STANDARD DEVIATION (σ) IS ESTIMATED FROM THE OBSERVED DATA

This procedure is the same as that described in Section 7.2. Use the values given in that section.

7.5 PAIRED SAMPLE *t*-TEST REQUIREMENTS

When performing the paired sample *t*-test, the estimate of the standard deviation of the difference will result from the paired sample. In many cases, a preliminary test will provide an estimate of σ from s, and this will enable us to calculate the required number of pairs for the necessary precision sought in the test. When performing the one-tail or two-tail test, we must determine the difference in standard deviation units (*d*).

$$d = \left| \frac{\mu_1 - \mu_2}{s} \right|$$

where

μ_1 = The mean value of population 1

μ_2 = The mean value of population 2

s = Sample standard deviation

The required sample size for a two-tail test is calculated by

$$n = \frac{(Z_{1-\alpha/2} + Z_{1-\beta})^2}{d^2} + \text{constant from Table 7.1}$$

where

$Z_{1-\alpha/2}$ = The normal distribution value for a given confidence level (see Appendix D, Selected Double-Sided Normal Distribution Probability Points)

$Z_{1-\beta}$ = The normal distribution value for a given confidence level (see Appendix C, Selected Single-Sided Normal Distribution Probability Points)

Example: We believe that a heat treatment process will change the size of a part by 0.024 or less. A preliminary sample of part sizes measured before and after heat treatment estimated the standard deviation of the paired differences to be .06. How large a sample is required to detect a difference in the averages of the paired values as low as .024 with an α of .05 and a β of .05?

$$d = \left| \frac{\mu_1 - \mu_2}{s} \right| = \left| \frac{0.024}{0.06} \right| = 0.40$$

$$n = \frac{(Z_{1-\alpha/2} + Z_{1-\beta})^2}{d^2} = \frac{(1.960 + 1.645)^2}{(0.40)^2} + 2 = 83.23$$

rounded up to the next integer is 84.

If the direction of the difference is important ($\mu_1 < \mu_2$ or $\mu_1 > \mu_2$), we use a one-tail test. The required sample size for a one-tail test is calculated by

$$n = \frac{(Z_{1-\alpha} + Z_{1-\beta})^2}{d^2} + \text{constant from Table 7.1}$$

where

$Z_{1-\alpha}$ = The normal distribution value for a given confidence level (see Appendix C, Selected Single-Sided Normal Distribution Probability Points)

$Z_{1-\beta}$ = The normal distribution value for a given confidence level (see Appendix C, Selected Single-Sided Normal Distribution Probability Points)

Example: If, instead of a two-tail test, it is important that the part size does not increase by 0.024 with all other parameters being the same, what is the number of paired samples required to detect a one-directional shift of 0.024?

$$d = \left|\frac{\mu_1 - \mu_2}{s}\right| = \left|\frac{0.024}{0.06}\right| = 0.40$$

$$n = \frac{(Z_{1-\alpha} + Z_{1-\beta})^2}{d^2} = \frac{(1.645 + 1.645)^2}{(0.40)^2} + 2 = 69.65$$

rounded up to the next integer is 70.

7.6 SAMPLE SIZE REQUIRED FOR CHI-SQUARE TEST OF OBSERVED VARIANCE TO A HYPOTHESIZED VARIANCE

When performing the chi-square test, there are two conditions of interest for the alternative hypothesis. These alternatives are:

$$\text{Case 1} \quad \sigma_n^2 > \sigma_c^2 \quad \text{and Case 2} \quad \sigma_n^2 < \sigma_c^2$$

where

σ_c^2 = Current of hypothesized variance

σ_n^2 = New or observed variance from sample data

When performing a chi-square test in case 1, we must specify the differences in variances we wish to detect at a given β risk. We will divide this larger standard deviation by the hypothesized standard deviation to determine the R ratio. Note that we use the standard deviations rather than the variance to calculate the ratio:

$$R = \frac{\sigma n}{\sigma c}$$

where

σn = Current of hypothesized standard deviation

σc = New or observed standard deviation from sample data

We calculate the sample size for a given R, α, and β as

$$n = 1 + .5 \left(\frac{Z_{1-\alpha} + (R * Z_{1-\beta})}{R - 1} \right)^2$$

where

$Z_{1-\alpha}$ = The normal distribution value for a given confidence level (see Appendix C, Selected Single-Sided Normal Distribution Probability Points)

$Z_{1-\beta}$ = The normal distribution value for a given confidence level (see Appendix C, Selected Single-Sided Normal Distribution Probability Points)

Example: A process is operating with a standard deviation of 0.0012. A new lower-cost process is proposed, but it will not be considered acceptable if the variability increases by more than 25%. Using an α of 0.05 and a β of 0.10 in detecting an increase to a standard deviation of 0.0015, calculate the sample size necessary to perform the chi-square test:

$$R = \frac{\sigma n}{\sigma c} = \frac{0.0015}{0.0012} = 1.25$$

$$n = 1 + .5 \left(\frac{Z_{1-\alpha} + (R * Z_{1-\beta})}{R - 1} \right)^2 = 1 + .5 \left(\frac{1.645 + (1.25 * 1.282)}{1.25 - 1} \right)^2 = 85.37$$

rounded up to the next integer is 86.

The sample size necessary to perform the chi-square test to determine if the new variance is greater than the current variance by a difference of 25% (case 1) with the given risks is 86.

When testing that the new variance is less than the current variance, the R ratio is calculated and will be less than 1. Calculating for case 2 use the following formula:

$$n = 1 + .5 \left(\frac{Z_{1-\alpha} + (R * Z_{1-\beta})}{1 - R} \right)^2$$

where

$Z_{1-\alpha}$ = The normal distribution value for a given confidence level (see Appendix C, Selected Single-Sided Normal Distribution Probability Points)

$Z_{1-\beta}$ = The normal distribution value for a given confidence level (see Appendix C, Selected Single-Sided Normal Distribution Probability Points)

Example: If in the previous example the new process were more expensive and would be adopted if the decrease in variability were 25%, and all other risks remain the same, we get

$$R = \frac{\sigma n}{\sigma c} = \frac{0.0009}{0.0012} = 0.75$$

$$n = 1 + .5 \left(\frac{Z_{1-\alpha} + (R * Z_{1-\beta})}{1 - R} \right)^2 = 1 + .5 \left(\frac{1.645 + (0.75 * 1.282)}{1 - 0.75} \right)^2 = 55.35$$

rounded up to the next integer is 56.

The sample size necessary to perform the chi-square test to determine if the new variance is less than the current variance by a difference of 25% (case 2) with the given risks is 56.

7.7 SAMPLE SIZE REQUIRED FOR *F*-TEST OF TWO OBSERVED VARIANCES

The F-test for testing the difference in two observed sample variances may be of interest in detecting a specific difference between two variances at a given α and β risk. For this method to be used, the two process sample sizes must be the same.

We will divide this larger standard deviation by the hypothesized standard deviation to determine the R ratio. Note that we used the standard deviations rather than the variance to calculate the ratio:

$$R = \frac{\sigma_1}{\sigma_2}$$

where

$$\sigma_1 > \sigma_2$$

where

σ_1 = Standard deviation 1

σ_2 = Standard deviation 2

$$n = 2 + \left(\frac{Z_{1-\alpha} + Z_{1-\beta}}{\ln(R)} \right)^2$$

where

$Z_{1-\alpha}$ = The normal distribution value for a given confidence level (see Appendix C, Selected Single-Sided Normal Distribution Probability Points)

$Z_{1-\beta}$ = The normal distribution value for a given confidence level (see Appendix C, Selected Single-Sided Normal Distribution Probability Points)

Example: Two processes are combined into one process stream. If the standard deviations of the processes differ by more than 30% ($\sigma_1 > 1.3\sigma$), the effects on the downstream variation are detrimental. Determine the sample sizes (which will be the same) necessary to detect a difference of this magnitude with an α risk of 0.05 and a β risk of 0.10. Note that we set the R value to 1.3 because we are looking for a difference of 30% or larger.

$$n = 2 + \left(\frac{1.960 + 1.645)}{\ln(1.3)} \right)^2 = 191.3$$

rounded up to the next integer is 192.

The sample size needed for each process is 192.

8

Hypothesis Testing for a Difference in Means

This chapter will focus on hypothesis testing for differences in means. In general, sample averages, counts, and proportions will have different values even if they are drawn from a single population. This section will demonstrate methods allowing the user to determine (at the appropriate confidence level) whether the differences noted in the samples are due to sampling variations or to real differences in the population average, count, or proportion.

8.1 TESTING A SAMPLE MEAN VERSUS A HYPOTHESIZED MEAN WHEN THE STANDARD DEVIATION (σ) IS KNOWN

When the standard deviation is known (or can be assumed), the distribution of averages drawn from a population will be distributed normally with a standard deviation of $\sigma / \sqrt{n}$, best used when the sample is > 30.

Using this information, we can develop hypothesis tests to determine the location (mean) of a population compared to a hypothesized or historical population mean.

The reference distribution of the test statistic is the normal distribution, and the formula is

$$Z = \frac{\bar{X} - \mu}{\sigma / \sqrt{n}}$$

where

$\bar{X}$ = Sample mean

σ = Population standard deviation

n = Number of samples

μ = Population mean

$Z_{\alpha =}$ The normal distribution value for a given confidence level (see Appendix C, Selected Single-Sided Normal Distribution Probability Points)

$Z_{\alpha/2}$ = The normal distribution value for a given confidence level (see
Appendix D, Selected Double-Sided Normal Distribution Probability
Points)

Example: To ensure that a new plant has the required capacity to meet demand, it is
necessary to maintain the holding time in the bulk tanks to an average of 180 minutes.
A similar process is being used at another facility. Using that process as the reference,
test whether the holding process averages 180 minutes at the 5% α risk level. The sample
size is 10, with an average of 184.6 minutes. The current process has an average holding
time of 180 minutes with a standard deviation of 3:

$H_0 : \mu = 180$ (null hypothesis two-tail test)

$H_A : \mu \neq 180$ (alternative hypothesis two-tail test)

$Z_\alpha = \pm 1.960$

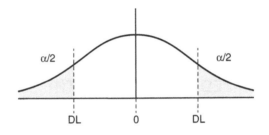

Z calculated		**Z critical**
$Z = \dfrac{184.6 - 180}{3 / \sqrt{10}} = 4.85$	>	± 1.960

Since Z calculated is greater than Z critical at the 0.05 α (alpha) level, there is sufficient
evidence to reject the null hypothesis $H_0 : \mu = 180$ in favor of the alternative hypothesis
that $H_1 : \mu \neq 180$.

In the previous example it was important to maintain the holding time to an average of
180 minutes to keep the material from premature curing prior to further use.

If the requirement were an average of 180 minutes or less, we would state the null
and alternative hypotheses in a different manner, and the result would be a one-sided test
with all the risk in the right tail of the distribution.

We would state the hypotheses as

$H_0 : \mu \leq 180$ (null hypothesis one-tail test)

$H_A : \mu > 180$ (alternative hypothesis one-tail test)

$Z_\alpha = 1.645$

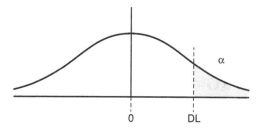

Z calculated	**Z critical**

$$Z = \frac{184.6 - 180}{3 / \sqrt{10}} = 4.85 \quad > \quad 1.645$$

Since Z calculated is greater than Z critical at the 0.05 α (alpha) level, there is sufficient evidence to reject the null hypothesis $H_0 : \mu \leq 180$ in favor of the alternative hypothesis that $H_A : \mu > 180$.

If the requirement were an average of 180 minutes or more, we would state the null and alternative hypotheses in a different manner, and the result would be a one-sided test with all the risk in the left tail of the distribution.

We would state the hypotheses as

$H_0 : \mu \geq 180$ (null hypothesis one-tail test)

$H_A : \mu < 180$ (alternative hypothesis one-tail test)

$Z_\alpha = -1.645$

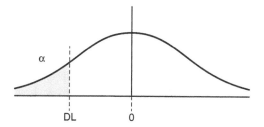

Z calculated	**Z critical**

$$Z = \frac{184.6 - 180}{3 / \sqrt{10}} = 4.85 \quad > \quad 1.645$$

Since Z calculated is less than Z critical at the 0.05 α (alpha) level, there is insufficient evidence to reject the null hypothesis that $H_0 : \mu \geq 180$.

8.2 TESTING A SAMPLE MEAN VERSUS A HYPOTHESIZED MEAN WHEN THE STANDARD DEVIATION (σ) IS ESTIMATED FROM THE SAMPLE DATA

When the standard deviation is unknown and must be estimated from a sample drawn from the population, we use the *t*-distribution to test the observed mean against a hypothesized mean, best used when the sample is < 30. One of the assumptions of this test is that the sample is drawn from a population that is normally distributed. The formula used is:

$$t = \frac{\bar{X} - \mu}{s / \sqrt{n}}$$

where

   $\bar{X}$ = Sample mean

   s = Sample standard deviation

   n = Number of samples

   μ = Population mean

   $t_{\alpha/2, n-1}$ = Double-sided test

   $t_{\alpha, n-1}$ = Single-sided test

   Percentage points of the Student's *t*-distribution value for a given confidence level for (n–1) degrees of freedom (see Appendix E, Percentage Points of the Student's *t*-Distribution)

Example: To ensure that a new plant has the required capacity to meet demand, it is necessary to maintain the holding time in the bulk tanks to an average of 180 minutes. A similar process is being used at another facility. Using that process as the reference, test whether the holding process averages 180 minutes at the 5% α risk level. The sample size is 10, with an average of 184.6 minutes with a sample standard deviation of 2.547. The current process has an average holding time of 180 minutes.

   $H_0 : \mu = 180$ (null hypothesis two-tail test)

   $H_A : \mu \neq 180$ (alternative hypothesis two-tail test)

   $t_{\alpha/2, n-1} = \pm 2.262$

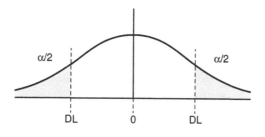

t calculated	*t* critical

$$t = \frac{184.6 - 180}{2.547 / \sqrt{10}} = 5.771 \quad > \quad \pm 2.262$$

Since *t* calculated is greater than *t* critical at the 0.05 α (alpha) level, there is sufficient evidence to reject the null hypothesis $H_0 : \mu = 180$ in favor of the alternative hypothesis that $H_A : \mu \neq 180$.

In the previous example, it was important to maintain the holding time to an average of 180 minutes to keep the material from premature curing prior to further use.

If the requirement were an average of 180 minutes or less, we would state the null and alternative hypotheses in a different manner, and the result would be a one-sided test with all the risk in the right tail of the distribution.

We would state the hypotheses as

$H_0 : \mu \leq 180$ (null hypothesis one-tail test)

$H_A : \mu > 180$ (alternative hypothesis one-tail test)

$t_{\alpha, n-1} = 1.833$

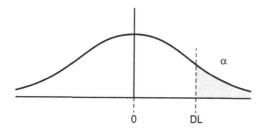

t calculated	*t* critical

$$t = \frac{184.6 - 180}{2.547 / \sqrt{10}} = 5.771 \quad > \quad 1.833$$

Since *t* calculated is greater than *t* critical at the 0.05 α (alpha) level, there is sufficient evidence to reject the null hypothesis $H_0 : \mu \leq 180$ in favor of the alternative hypothesis that $H_A : \mu > 180$.

If the requirement were an average of 180 minutes or more, we would state the null and alternative hypotheses in a different manner, and the result would be a one-sided test with all the risk in the left tail of the distribution.

We would state the hypotheses as

$H_0 : \mu \geq 180$ (null hypothesis one-tail test)

$H_A : \mu < 180$ (alternative hypothesis one-tail test)

$t_{\alpha, n-1} = 1.833$

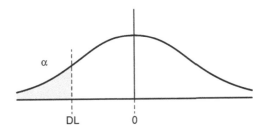

t **calculated**	*t* **critical**

$$t = \frac{184.6 - 180}{2.547 / \sqrt{10}} = 5.771 \quad > \quad -1.833$$

Since *t* calculated is less than *t* critical at the 0.05 α (alpha) level, there is insufficient evidence to reject the null hypothesis that $H_0 : \mu \geq 180$.

8.3 TESTING FOR A DIFFERENCE IN TWO POPULATION MEANS—STANDARD DEVIATIONS (σ) KNOWN

When using samples drawn from two populations to test for differences in their mean values when the standard deviation is known, the reference distribution is the normal distribution, and we use the following formula:

$$Z = \frac{\bar{X}_1 - \bar{X}_2}{\sqrt{s_1^2 / n_1 + s_2^2 / n_2}}$$

where

$\bar{X}_1$ = Sample mean of population 1

$\bar{X}_2$ = Sample mean of population 2

s_1 = Standard deviation of population 1

s_2 = Standard deviation of population 2

n_1 = Number of samples drawn from population 1

n_2 = Number of samples drawn from population 2

Z_α = The normal distribution value for a given confidence level (see Appendix C, Selected Single-Sided Normal Distribution Probability Points)

$Z_{\alpha/2}$ = The normal distribution value for a given confidence level (see Appendix D, Selected Double-Sided Normal Distribution Probability Points)

Example: We wish to compare the two processes to see if process 2 is the same as process 1 at 5% α risk level.

Process 1	Process 2
85.2	89.0
87.3	89.4
95.2	90.8
80.8	84.3
84.8	88.2
	88.1
$\overline{X} = 86.12$	$\overline{X} = 88.3$
$n = 5$	$n = 6$
$\sigma = 4.27$	$\sigma = 2.19$

$H_0 : \mu_1 = \mu_2$ (null hypothesis two-tail test)

$H_A : \mu_1 \neq \mu_2$ (alternative hypothesis two-tail test)

$Z_\alpha = \pm 1.960$

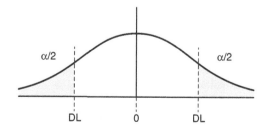

Z calculated	**Z critical**

$$Z = \frac{\overline{X}_1 - \overline{X}_2}{\sqrt{s_1^2/n_1 + s_2^2/n_2}} = \frac{86.12 - 88.3}{\sqrt{4.27^2/5 + 2.19^2/6}} = -1.034 \quad > \quad \pm 1.960$$

Since Z calculated is less than Z critical at the 0.05 α (alpha) level, there is insufficient evidence to reject the null hypothesis that $H_0 : \mu_1 = \mu_2$.

If the requirement were that process 1 is equal to or less than process 2, we would state the null and alternative hypotheses in a different manner, and the result would be a one-sided test with all the risk in the right tail of the distribution.

We would state the hypotheses as

$H_0 : \mu_1 \leq \mu_2$ (null hypothesis one-tail test)

$H_A : \mu_1 > \mu_2$ (alternative hypothesis one-tail test)

$Z_\alpha = 1.645$

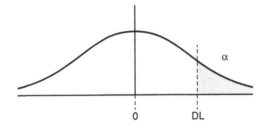

Z calculated	**Z critical**

$$Z = \frac{\overline{X}_1 - \overline{X}_2}{\sqrt{s_1^2/n_1 + s_2^2/n_2}} = \frac{86.12 - 88.3}{\sqrt{4.27^2/5 + 2.19^2/6}} = -1.034 \quad < \quad 1.645$$

Since Z calculated is less than Z critical at the 0.05 α (alpha) level, there is insufficient evidence to reject the null hypothesis that $H_0 : \mu_1 \leq \mu_2$.

If the requirement were that process 1 is equal to or greater than process 2, we would state the null and alternative hypothesis in a different manner, and the result would be a one-sided test with all the risk in the left tail of the distribution.

We would state the hypotheses as

$H_0 : \mu_1 \geq \mu_2$ (null hypothesis one-tail test)

$H_A : \mu_1 < \mu_2$ (alternative hypothesis one-tail test)

$Z_\alpha = -1.645$

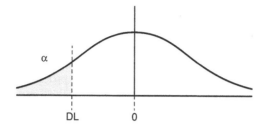

Z calculated		*Z* critical

$$Z = \frac{\bar{X}_1 - \bar{X}_2}{\sqrt{s_1^2/n_1 + s_2^2/n_2}} = \frac{86.12 - 88.3}{\sqrt{4.27^2/5 + 2.19^2/6}} = -1.034 \quad < \quad -1.645$$

Since *Z* calculated is less than *Z* critical at the 0.05 α (alpha) level, there is insufficient evidence to reject the null hypothesis that $H_0 : \mu_1 \geq \mu_2$.

8.4 TESTING A SAMPLE MEAN VERSUS A HYPOTHESIZED MEAN WHEN THE STANDARD DEVIATION (σ) IS ESTIMATED FROM THE SAMPLE DATA

When the population standard deviation must be estimated from samples drawn from two normally distributed populations to test for differences in their mean values, the appropriate reference distribution is the *t*-distribution, and we use this equation:

$$t = \frac{\bar{X}_1 - \bar{X}_2}{s_p\sqrt{1/n_1 + 1/n_2}}$$

$$s_p = \sqrt{\frac{(n_1 - 1)(s_1^2) + (n_2 - 1)(s_2^2)}{n_1 + n_2 - 2}}$$

where

$\bar{X}_1$ = Sample mean of population 1

$\bar{X}_2$ = Sample mean of population 2

s_1 = Standard deviation of population 1

s_2 = Standard deviation of population 2

n_1 = Number of samples drawn from population 1

n_2 = Number of samples drawn from population 2

$t_{\alpha/2, n_1 + n_2 - 2}$ = Double-sided test

$t_{\alpha, n_1 + n_2 - 2}$ = Single-sided test

Percentage points of the Student's *t*-distribution value for a given confidence level for $(n-1)$ degrees of freedom (see Appendix E, Percentage Points of the Student's *t*-Distribution)

Example: We wish to compare two processes to see if process 2 is the same as process 1 at 5% α risk level.

Process 1	Process 2
85.2	89.0
87.3	89.4
95.2	90.8
80.8	84.3
84.8	88.2
	88.1
$\bar{X} = 86.12$	$\bar{X} = 88.3$
$n = 5$	$n = 6$
$\sigma = 4.27$	$\sigma = 2.19$

$H_0 : \mu_1 = \mu_2$ (null hypothesis two-tail test)

$H_A : \mu_1 \neq \mu_2$ (alternative hypothesis two-tail test)

$t_{\alpha/2, n_1 + n_2 - 2} = \pm 2.262$

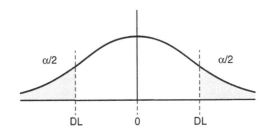

$$S_p = \sqrt{\frac{(n_1 - 1)(s_1^2) + (n_2 - 1)(s_2^2)}{n_1 + n_2 - 2}} = \sqrt{\frac{(5-1)(4.27^2) + (6-1)(2.19^2)}{5 + 6 - 2}} = 3.281$$

t calculated	**t critical**

$$t = \frac{\bar{X}_1 - \bar{X}_2}{S_p\sqrt{1/n_1 + 1/n_2}} = \frac{86.12 - 88.3}{(3.281)\sqrt{1/5 + 1/6}} = -1.097 \quad < \quad \pm 2.262$$

Since *t* calculated is less than *t* critical at the 0.05 α (alpha) level, there is insufficient evidence to reject the null hypothesis that $H_0 : \mu_1 = \mu_2$.

If the requirement were that process 1 is equal to or less than process 2, we would state the null and alternative hypotheses in a different manner, and the result would be a one-sided test with all the risk in the right tail of the distribution.

We would state the hypotheses as

$H_0 : \mu_1 \leq \mu_2$ (null hypothesis one-tail test)

$H_A : \mu_1 > \mu_2$ (alternative hypothesis one-tail test)

$t_{\alpha, n_1 + n_2 - 2} = 1.833$

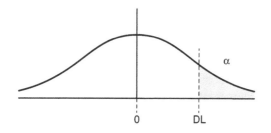

t **calculated**	*t* **critical**

$$t = \frac{\bar{X}_1 - \bar{X}_2}{s_p\sqrt{1/n_1 + 1/n_2}} = \frac{86.12 - 88.3}{(3.281)\sqrt{1/5 + 1/6}} = -1.097 \quad < \quad 1.833$$

Since *t* calculated is less than *t* critical at the 0.05 α (alpha) level, there is insufficient evidence to reject the null hypothesis that $H_0 : \mu_1 \leq \mu_2$.

If the requirement were that process 1 is equal to or greater than process 2, we would state the null and alternative hypotheses in a different manner, and the result would be a one-sided test with all the risk in the left tail of the distribution.

We would state the hypotheses as

$H_0 : \mu_1 \geq \mu_2$ (null hypothesis one-tail test)

$H_A : \mu_1 < \mu_2$ (alternative hypothesis one-tail test)

$t_{\alpha, n_1 + n_2 - 2} = 1.833$

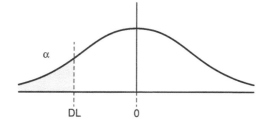

t calculated	*t* critical

$$t = \frac{\bar{X}_1 - \bar{X}_2}{s_p\sqrt{1/n_1 + 1/n_2}} = \frac{86.12 - 88.3}{(3.281)\sqrt{1/5 + 1/6}} = -1.097 \quad > \quad -1.833$$

Since *t* calculated is less than *t* critical at the 0.05 α (alpha) level, there is insufficient evidence to reject the null hypothesis that $H_0 : \mu_1 \geq \mu_2$.

8.5 TESTING FOR A DIFFERENCE IN TWO POPULATION MEANS—STANDARD DEVIATIONS (σ) NOT KNOWN AND NOT ASSUMED EQUAL

As mentioned in Section 8.4, one assumption of the two-sample *t*-test is that of equal standard deviations (variances). When this assumption is not valid, we may use an alternative test outlined by Natrella (1966), the Aspin-Welch test. While it is similar to the two-sample *t*-test, there are some distinct differences, the first being the calculation of the degrees of freedom (df):

$$c = \frac{\dfrac{s_1^2}{n_1}}{\dfrac{s_1^2}{n_1} + \dfrac{s_2^2}{n_2}}$$

$$df = \frac{1}{\dfrac{c^2}{n_1 - 1} + \dfrac{(1-c)^2}{n_2 - 1}}$$

$$t = \frac{\bar{X}_1 - \bar{X}_2}{\sqrt{\dfrac{s_1^2}{n_1} + \dfrac{s_2^2}{n_2}}}$$

where

$\bar{X}_1$ = Sample mean of population 1

$\bar{X}_2$ = Sample mean of population 2

s_1 = Standard deviation of population 1

s_2 = Standard deviation of population 2

n_1 = Number of samples drawn from population 1

n_2 = Number of samples drawn from population 2

$t_{\alpha/2,df(Calculated)}$ = Double-sided test

$t_{\alpha,df(Calculated)}$ = Single-sided test

Percentage points of the Student's t-distribution value for a given confidence level for the calculated degrees of freedom (see Appendix E, Percentage Points of the Student's t-Distribution)

Example: Two machine tools are set up with similar fixturing and process parameters to determine if the averages of parts machined on the two machines are equivalent. Test to determine if the averages are equal at the 5% level of risk.

Machine 1	Machine 2
$\bar{X}$ = 201.8 mm	$\bar{X}$ = 203.5 mm
n = 12	n = 11
s = 4.4	s = 2.3

$H_0 : u_1 = \mu_2$ (null hypothesis two-tail test)

$H_A : u_1 \neq \mu_2$ (alternative hypothesis two-tail test)

$t_{\alpha/2,df(Calculated)} = \pm 2.120$

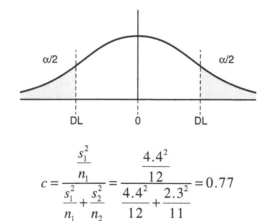

$$c = \frac{\dfrac{s_1^2}{n_1}}{\dfrac{s_1^2}{n_1} + \dfrac{s_2^2}{n_2}} = \frac{\dfrac{4.4^2}{12}}{\dfrac{4.4^2}{12} + \dfrac{2.3^2}{11}} = 0.77$$

$$df = \frac{1}{\dfrac{c^2}{n_1 - 1} + \dfrac{(1-c)^2}{n_2 - 1}} = \frac{1}{\dfrac{0.77^2}{12 - 1} + \dfrac{(1-0.77)^2}{11 - 1}} = 16.89 \text{ (rounded down to 16)}$$

| **_t_ calculated** | **_t_ critical** |

$$t = \frac{\bar{X}_1 - \bar{X}_2}{\sqrt{\dfrac{s_1^2}{n_1} + \dfrac{s_2^2}{n_2}}} = \frac{201.8 - 203.5}{\sqrt{\dfrac{4.4^2}{12} + \dfrac{2.3^2}{11}}} = -1.174 \quad < \quad \pm 2.120$$

Since t calculated is less than t critical at the 0.05 α (alpha) level, there is insufficient evidence to reject the null hypothesis that $H_0 : \mu_1 = \mu_2$.

If the requirement were that machine 1 is equal to or less than machine 2, we would state the null and alternative hypotheses in a different manner, and the result would be a one-sided test with all the risk in the right tail of the distribution.

We would state the hypotheses as

$H_0 : \mu_1 \leq \mu_2$ (null hypothesis one-tail test)

$H_A : \mu_1 > \mu_2$ (alternative hypothesis one-tail test)

$t_{\alpha,df(\text{Calculated})} = 1.746$

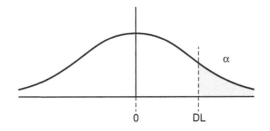

| **_t_ calculated** | **_t_ critical** |

$$t = \frac{\bar{X}_1 - \bar{X}_2}{s_p \sqrt{1/n_1 + 1/n_2}} = \frac{86.12 - 88.3}{(3.281)\sqrt{1/5 + 1/6}} = -1.097 \quad < \quad 1.746$$

Since t calculated is less than t critical at the 0.05 α (alpha) level, there is insufficient evidence to reject the null hypothesis that $H_0 : \mu_1 \leq \mu_2$.

If the requirement were that machine 1 is equal to or greater than machine 2, we would state the null and alternative hypotheses in a different manner, and the result would be a one-sided test with all the risk in the left tail of the distribution.

We would state the hypotheses as

$H_0 : \mu_1 \geq \mu_2$ (null hypothesis one-tail test)

$H_A : \mu_1 < \mu_2$ (alternative hypothesis one-tail test)

$t_{\alpha,df(\text{Calculated})} = -1.746$

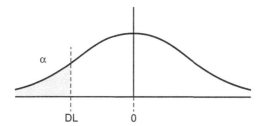

t calculated   **t critical**

$$t = \frac{\bar{X}_1 - \bar{X}_2}{s_p \sqrt{1/n_1 + 1/n_2}} = \frac{86.12 - 88.3}{(3.281)\sqrt{1/5 + 1/6}} = -1.097 \quad < \quad -1.746$$

Since *t* calculated is less than *t* critical at the 0.05 α (alpha) level, there is insufficient evidence to reject the null hypothesis that $H_0 : \mu_1 \geq \mu_2$.

8.6 TESTING FOR DIFFERENCES IN MEANS OF PAIRED SAMPLES

When testing for the difference of two means before and after a treatment, we can utilize the paired-sample *t*-test. This test could determine if there is a difference in a tensile test before and after a chemical treatment. We use the following formula:

$$t = \frac{\bar{d}}{s/\sqrt{n}}$$

where

$\bar{d}$ = Average difference of the paired samples

s = Standard deviation of the differences

n = Number of pairs

df = Number of pairs minus 1

$t_{\alpha/2,n-1}$ = Double-sided test

$t_{\alpha,n-1}$ = Single-sided test

Percentage points of the Student's *t*-distribution value for a given confidence level for the calculated degrees of freedom (see Appendix E, Percentage Points of the Student's *t*-Distribution)

Example: We test 10 employees to determine the time (in minutes) to complete a given task using written instructions. We then give the same employees video-based training to determine if such training enables them to perform the task in less time. At the 95%

confidence level, test the hypothesis that there is no difference between the time required to complete the task prior to the video training and after the video training.

Sample	Before video	After video	Difference
1	27	18	−9
2	17	23	6
3	22	20	−2
4	26	23	−3
5	26	28	2
6	30	31	1
7	20	20	0
8	20	17	−3
9	21	19	−2
10	27	24	−3
			$\bar{d} = -1.3$
			$s = 3.945$
			$n = 10$

$H_0 : \mu_1 = \mu_2$ (null hypothesis two-tail test)

$H_A : \mu_1 \neq \mu_2$ (alternative hypothesis two-tail test)

$t_{\alpha/2, n-1} = \pm 2.262$

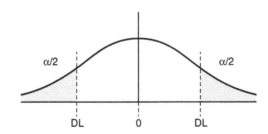

t calculated   **t critical**

$$t = \frac{\bar{d}}{s / \sqrt{n}} = \frac{-1.3}{(3.945)\sqrt{10}} = -1.042 \quad < \quad \pm 2.262$$

Since *t* calculated is less than *t* critical at the 0.05 α (alpha) level, there is insufficient evidence to reject the null hypothesis that $H_0 : \mu_1 = \mu_2$.

If we wish to verify that the results before training are equal to or less than the results after training, we would state the null and alternative hypotheses in a different manner, and the result would be a one-sided test with all the risk in the right tail of the distribution.

We would state the hypotheses as

$H_0 : \mu_1 \leq \mu_2$ (null hypothesis one-tail test)

$H_A : \mu_1 > \mu_2$ (alternative hypothesis one-tail test)

$t_{\alpha,n-1} = 1.833$

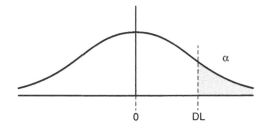

t calculated	*t* critical

$$t = \frac{\overline{d}}{s/\sqrt{n}} = \frac{-1.3}{(3.945)\sqrt{10}} = -1.042 \quad < \quad 1.833$$

Since *t* calculated is less than *t* critical at the 0.05 α (alpha) level, there is insufficient evidence to reject the null hypothesis that $H_0 : \mu_1 \leq \mu_2$.

If we wish to verify that the results before training are equal to or greater than the results after training, we would state the null and alternative hypotheses in a different manner, and the result would be a one-sided test with all the risk in the left tail of the distribution.

$H_0 : \mu_1 \geq \mu_2$ (null hypothesis one-tail test)

$H_A : \mu_1 < \mu_2$ (alternative hypothesis one-tail test)

$t_{\alpha,n-1} = -1.833$

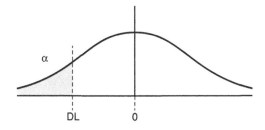

t calculated	*t* critical

$$t = \frac{\overline{d}}{s/\sqrt{n}} = \frac{-1.3}{(3.945)\sqrt{10}} = -1.042 \quad < \quad -1.833$$

Since *t* calculated is less than *t* critical at the 0.05 α (alpha) level, there is insufficient evidence to reject the null hypothesis that $H_0 : \mu_1 \geq \mu_2$.

8.7 TESTING FOR DIFFERENCES IN TWO PROPORTIONS

When testing for differences in proportions using large sample sizes, and np and $n(1-p)$ are both greater than 5, we use the following equations:

$$p' = \frac{(n_1)(p_1) + (n_2)(p_2)}{n_1 + n_2}$$

$$^s p_1 - p_2 = \sqrt{p'(1-p')(1/n_1 + 1/n_2)}$$

$$Z = \frac{p_1 - p_2}{^s p_1 - p_2}$$

where

p_1 = Sample proportion 1

p_2 = Sample proportion 2

n_1 = Number of samples for proportion 1

n_2 = Number of samples for proportion 2

Z_α = The normal distribution value for a given confidence level (see Appendix C, Selected Single-Sided Normal Distribution Probability Points)

$Z_{\alpha/2}$ = The normal distribution value for a given confidence level (see Appendix D, Selected Double-Sided Normal Distribution Probability Points)

Example: We test two competing emerging-technology processes to determine if the fraction of nonconforming product manufactured by each process is equivalent. We draw a large sample from each process. Determine if the fraction nonconforming for each process is the same at the 95% confidence level.

Process 1	Process 2
$n = 300$	$n = 350$
$d = 12$	$d = 21$
$p = 21/300 = 0.04$	$p = 21/350 = 0.06$

$H_0 : \mu_1 = \mu_2$ (null hypothesis two-tail test)

$H_A : \mu_1 \neq \mu_2$ (alternative hypothesis two-tail test)

$Z_\alpha = \pm 1.960$

$$p' = \frac{(n_1)(p_1) + (n_2)(p_2)}{n_1 + n_2} = \frac{(300)(0.04) + (350)(0.06)}{300 + 350} = 0.051$$

$$Z = \frac{p_1 - p_2}{s_{p_1 - p_2}} = \frac{0.04 - 0.06}{0.017} = -1.176$$

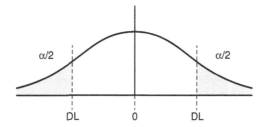

Z calculated	**Z critical**

$$Z = \frac{p_1 - p_2}{s_{p_1 - p_2}} = \frac{0.04 - 0.06}{0.017} = -1.176 \quad < \quad \pm 1.960$$

Since Z calculated is less than Z critical at the 0.05 α (alpha) level, there is insufficient evidence to reject the null hypothesis that $H_0 : \mu_1 = \mu_2$.

If the requirement were that process 1 is equal to or less than process 2, we would state the null and alternative hypotheses in a different manner, and the result would be a one-sided test with all the risk in the right tail of the distribution.

We would state the hypotheses as

$H_0 : \mu_1 \leq \mu_2$ (null hypothesis one-tail test)

$H_A : \mu_1 > \mu_2$ (alternative hypothesis one-tail test)

$Z_\alpha = 1.645$

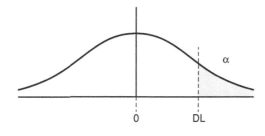

Z calculated	**Z critical**

$$Z = \frac{p_1 - p_2}{s_{p_1 - p_2}} = \frac{0.04 - 0.06}{0.017} = -1.176 \quad < \quad 1.645$$

Since Z calculated is less than Z critical at the 0.05 α (alpha) level, there is insufficient evidence to reject the null hypothesis that $H_0 : \mu_1 \leq \mu_2$.

If the requirement were that process 1 is equal to or greater than process 2, we would state the null and alternative hypotheses in a different manner, and the result would be a one-sided test with all the risk in the left tail of the distribution.

We would state the hypotheses as

$H_0 : \mu_1 \geq \mu_2$ (null hypothesis one-tail test)

$H_A : \mu_1 < \mu_2$ (alternative hypothesis one-tail test)

$Z_\alpha = 1.645$

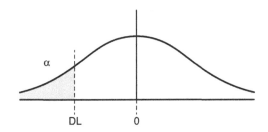

Z calculated	**Z critical**

$$Z = \frac{p_1 - p_2}{s_{p_1 - p_2}} = \frac{0.04 - 0.06}{0.017} = -1.176 \quad > \quad -1.645$$

Since Z calculated is less than Z critical at the 0.05 α (alpha) level, there is insufficient evidence to reject the null hypothesis that $H_0 : \mu_1 \geq \mu_2$.

8.8 TESTING FOR DIFFERENCES IN COUNT DATA— EQUAL SAMPLE SIZES

When testing for differences in count data, such as nonconformances per unit, when the sample size is relatively large and equal, we can perform a test of hypothesis using the normal distribution approximation:

$$Z = \frac{|Y_1 - Y_2| - 0.5}{\sqrt{Y_1 + Y_2}}$$

where

Y_1 = Number of occurrences in sample 1

Y_2 = Number of occurrences in sample 2

Z_α = The normal distribution value for a given confidence level (see Appendix C, Selected Single-Sided Normal Distribution Probability Points)

$Z_{\alpha/2}$ = The normal distribution value for a given confidence level (see Appendix D, Selected Double-Sided Normal Distribution Probability Points)

Example: A check of 10 assemblies noted 260 nonconformances. After investigation and corrective action, a check of 10 assemblies noted 215 nonconformances. At the 95% confidence level, test the hypothesis that the average number of nonconformances per unit is the same before and after corrective action.

$H_0 : \mu_1 = \mu_2$ (null hypothesis two-tail test)

$H_A : \mu_1 \neq \mu_2$ (alternative hypothesis two-tail test)

$Z_{\alpha/2} = \pm1.96$

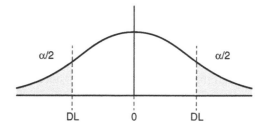

Z calculated   **Z critical**

$$Z = \frac{|Y_1 - Y_2| - 0.5}{\sqrt{Y_1 + Y_2}} = \frac{|260 - 215| - 0.5}{\sqrt{260 + 215}} = 2.042 \quad > \quad \pm1.960$$

Since Z calculated is greater than Z critical at the 0.05 α (alpha) level, there is sufficient evidence to reject the null hypothesis H_0: $\mu_1 = \mu_2$ in favor of the alternative hypothesis that $H_A : \mu_1 \neq \mu_2$.

If the requirement were that results before corrective action be equal to or less than the results after corrective action, we would state the null and alternative hypotheses in a different manner, and the result would be a one-sided test with all the risk in the right tail of the distribution.

We would state the hypotheses as

$H_0 : \mu_1 \leq \mu_2$ (null hypothesis one-tail test)

$H_A : \mu_1 > \mu_2$ (alternative hypothesis one-tail test)

$Z_\alpha = 1.645$

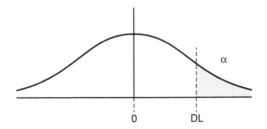

Z calculated		**Z critical**

$$Z = \frac{|Y_1 - Y_2| - 0.5}{\sqrt{Y_1 + Y_2}} = \frac{|260 - 215| - 0.5}{\sqrt{260 + 215}} = 2.042 \quad > \quad 1.645$$

Since Z calculated is greater than Z critical at the 0.05 α (alpha) level, there is sufficient evidence to reject the null hypothesis $H_0 : \mu_1 \geq \mu_2$ in favor of the alternative hypothesis that $H_A : \mu_1 > \mu_2$.

If the requirement were that results before corrective action be equal to or greater than the results after corrective action, we would state the null and alternative hypotheses in a different manner, and the result would be a one-sided test with all the risk in the left tail of the distribution.

We would state the hypotheses as

$H_0 : \mu_1 \geq \mu_2$ (null hypothesis one-tail test)

$H_A : \mu_1 < \mu_2$ (alternative hypothesis one-tail test)

$Z_\alpha = -1.645$

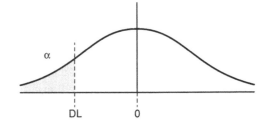

Z calculated	**Z critical**

$$Z = \frac{|Y_1 - Y_2| - 0.5}{\sqrt{Y_1 + Y_2}} = \frac{|260 - 215| - 0.5}{\sqrt{260 + 215}} = 2.042 \quad > \quad -1.645$$

Since Z calculated is less than Z critical at the 0.05 α (alpha) level, there is insufficient evidence to reject the null hypothesis that $H_0 : \mu_1 \geq \mu_2$.

8.9 TESTING FOR DIFFERENCES IN COUNT DATA— UNEQUAL SAMPLE SIZES

When testing for differences in count data, such as nonconformances per unit, when the sample size is relatively large and unequal, we can perform a test of hypothesis using the normal distribution approximation:

$$Z = \frac{n_2 Y_1 - n_1 Y_2}{(\sqrt{n_1 * n_2})(\sqrt{Y_1 + Y_2})}$$

Note the transposition of the sample sizes in the numerator of this equation.

where

Y_1 = Number of occurrences in sample 1

Y_2 = Number of occurrences in sample 2

n_1 = Number of units in sample 1

n_2 = Number of units in sample 2

Z_α = The normal distribution value for a given confidence level (see Appendix C, Selected Single-Sided Normal Distribution Probability Points)

$Z_{\alpha/2}$ = The normal distribution value for a given confidence level (see Appendix D, Selected Double-Sided Normal Distribution Probability Points)

Example: In a sample of 40 units, a total of 67 occurrences of a particular nonconformance were noted. After a design change to aid assembly, in a sample of 32 units, 31 occurrences of this nonconformance were noted. At the 95% confidence level, test the hypothesis that the average number of nonconformances per unit is the same before and after the design change.

$H_0 : \mu_1 = \mu_2$ (null hypothesis two-tail test)

$H_A : \mu_1 \neq \mu_2$ (alternative hypothesis two-tail test)

$Z_{\alpha/2} = \pm 1.96$

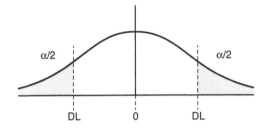

Z calculated   **Z critical**

$$Z = \frac{n_2 Y_1 - n_1 Y_2}{(\sqrt{n_1 * n_2})(\sqrt{Y_1 + Y_2})} = \frac{(32)(67) - (40)(31)}{(\sqrt{40 * 32})(\sqrt{67 + 31})} = 2.552 \quad > \quad \pm 1.960$$

Since Z calculated is greater than Z critical at the 0.05 α (alpha) level, there is sufficient evidence to reject the null hypothesis $H_0 : \mu_1 = \mu_2$ in favor of the alternative hypothesis that $H_A : \mu_1 \neq \mu_2$.

If the requirement were that results before corrective action are equal to or less than the results after corrective action, we would state the null and alternative hypotheses in a different manner, and the result would be a one-sided test with all the risk in the right tail of the distribution.

We would state the hypotheses as:

$H_0 : \mu_1 \leq \mu_2$ (null hypothesis one-tail test)

$H_A : \mu_1 > \mu_2$ (alternative hypothesis one-tail test)

$Z_\alpha = 1.645$

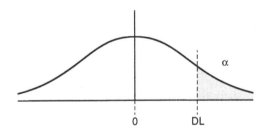

Z calculated   **Z critical**

$$Z = \frac{n_2 Y_1 - n_1 Y_2}{(\sqrt{n_1 * n_2})(\sqrt{Y_1 + Y_2})} = \frac{(32)(67) - (40)(31)}{(\sqrt{40 * 32})(\sqrt{67 + 31})} = 2.552 \quad > \quad 1.645$$

Since Z calculated is greater than Z critical at the 0.05 α (alpha) level, there is sufficient evidence to reject the null hypothesis $H_0 : \mu_1 \leq \mu_2$ in favor of the alternative hypothesis that $H_A : \mu_1 > \mu_2$.

If the requirement were that results before corrective action be equal to or greater than results after corrective action, we would state the null and alternative hypotheses in a different manner, and the result would be a one-sided test with all the risk in the left tail of the distribution.

We would state the hypotheses as:

$H_0 : \mu_1 \geq \mu_2$ (null hypothesis one-tail test)

$H_A : \mu_1 < \mu_2$ (alternative hypothesis one-tail test)

$Z_\alpha = -1.645$

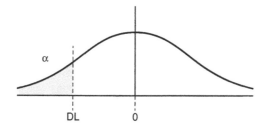

Z calculated	**Z critical**

$$Z = \frac{n_2 Y_1 - n_1 Y_2}{(\sqrt{n_1 * n_2})(\sqrt{Y_1 + Y_2})} = \frac{(32)(67) - (40)(31)}{(\sqrt{40 * 32})(\sqrt{67 + 31})} = 2.552 \quad > \quad -1.645$$

Since Z calculated is less than Z critical at the 0.05 α (alpha) level, there is insufficient evidence to reject the null hypothesis that $H_0 : \mu_1 \geq \mu_2$.

8.10 HYPOTHESIS TESTING FOR DIFFERENCES IN MEANS—CONFIDENCE INTERVAL APPROACH (STANDARD DEVIATIONS KNOWN)

Another approach to testing for differences in two means is the confidence interval approach. If the confidence interval calculation contains the value of zero, the hypothesis that the two means are equal is not rejected. If the confidence interval does not contain zero, the null hypothesis is rejected in favor of the alternative hypothesis.

$$\overline{X}_1 - \overline{X}_2 \pm Z_{\alpha/2} \sqrt{\frac{\sigma_1^2}{n_1} + \frac{\sigma_2^2}{n_2}}$$

where

$\overline{X}_1$ = Sample mean of population 1

$\bar{X}_2$ = Sample mean of population 2

σ_1 = Standard deviation of population 1

σ_2 = Standard deviation of population 2

n_1 = Number of samples drawn from population 1

n_2 = Number of samples drawn from population 2

$Z_{\alpha/2}$ = The normal distribution value for a given confidence level (see Appendix D, Selected Double-Sided Normal Distribution Probability Points)

Example: We wish to compare two processes to see if process 2 is the same as process 1 at 5% α risk level.

Process 1	Process 2
85.2	89.0
87.3	89.4
95.2	90.8
80.8	84.3
84.8	88.2
	88.1
$\bar{X}$ = 86.12	$\bar{X}$ = 88.3
$n = 5$	$n = 6$
$\sigma = 4.27$	$\sigma = 2.19$

$H_0 : \mu_1 = \mu_2$ (null hypothesis two-tail test)

$H_A : \mu_1 \neq \mu_2$ (alternative hypothesis two-tail test)

$Z_{\alpha} = \pm 1.960$

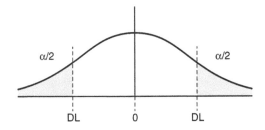

$$= \bar{X}_1 - \bar{X}_2 \pm Z_{\alpha/2} \sqrt{\frac{\sigma_1^2}{n_1} + \frac{\sigma_2^2}{n_2}}$$

$$= 86.12 - 88.3 \pm (1.960) \sqrt{\frac{4.13}{5} + \frac{2.5}{6}}$$

$$= -2.18 \pm 1.960(1.115)$$

$$= -2.18 \pm 2.593$$

$$= -4.77 \text{ to } 0.413$$

Since the confidence interval does not contain the value of zero at the 0.05 α (alpha) level, we reject the null hypothesis $H_0 : \mu_1 = \mu_2$ in favor of the alternative hypothesis $H_A : \mu_1 \neq \mu_2$.

8.11 HYPOTHESIS TESTING FOR DIFFERENCES IN MEANS—CONFIDENCE INTERVAL APPROACH (STANDARD DEVIATIONS NOT KNOWN BUT ASSUMED EQUAL)

When the standard deviations are unknown but assumed equal, the formula to calculate the confidence interval for the difference of the two means becomes:

$$s_p = \sqrt{\frac{(n_1 - 1)(s_1^2) + (n_2 - 1)(s_2^2)}{n_1 + n_2 - 2}}$$

$$\bar{X}_1 - \bar{X}_2 \pm t_{\alpha/2, n_1 + n_2 - 2} s_p \sqrt{1/n_1 + 1/n_2}$$

where

$\bar{X}_1$ = Sample mean of population 1

$\bar{X}_2$ = Sample mean of population 2

s_1 = Standard deviation of population 1

s_2 = Standard deviation of population 2

n_1 = Number of samples drawn from population 1

n_2 = Number of samples drawn from population 2

$t_{\alpha/2, n_1 + n_2 - 2}$ = Double-sided test

Percentage points of the Student's t-distribution value for a given confidence level for $(n - 1)$ degrees of freedom (see Appendix E, Percentage Points of the Student's t-Distribution)

Example: We wish to compare the two processes to see if process 2 is the same as process 1 at 5% α risk level.

Process 1	Process 2
85.2	89.0
87.3	89.4
95.2	90.8
80.8	84.3
84.8	88.2
	88.1
$\bar{X} = 86.12$	$\bar{X} = 88.3$
$n = 5$	$n = 6$
$\sigma = 4.27$	$\sigma = 2.19$

$H_0 : \mu_1 = \mu_2$ (null hypothesis two-tail test)

$H_A : \mu_1 \neq \mu_2$ (alternative hypothesis two-tail test)

$t_{\alpha/2, n_1 + n_2 - 2} = \pm 2.262$

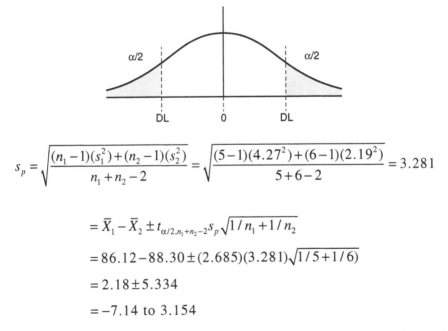

$$s_p = \sqrt{\frac{(n_1 - 1)(s_1^2) + (n_2 - 1)(s_2^2)}{n_1 + n_2 - 2}} = \sqrt{\frac{(5-1)(4.27^2) + (6-1)(2.19^2)}{5 + 6 - 2}} = 3.281$$

$$= \bar{X}_1 - \bar{X}_2 \pm t_{\alpha/2, n_1 + n_2 - 2} s_p \sqrt{1/n_1 + 1/n_2}$$

$$= 86.12 - 88.30 \pm (2.685)(3.281)\sqrt{1/5 + 1/6}$$

$$= 2.18 \pm 5.334$$

$$= -7.14 \text{ to } 3.154$$

Since the confidence interval does contain the value of zero at the 0.05 α (alpha) level, we fail to reject the null hypothesis that $H_0 : \mu_1 = \mu_2$.

9

Hypothesis Testing for a Difference in Variances

There are three common tests for variances: (1) testing to determine if a variance of a population as estimated from a sample equals a hypothesized or known variance, (2) testing to determine if two variances estimated from two samples could have come from populations having equal variances, and (3) testing for the equality of several variances, as in an *analysis of variance* (ANOVA).

9.1 TESTING A VARIANCE CALCULATED FROM A SAMPLE AGAINST A HYPOTHESIZED VARIANCE

To test if a variance calculated from sample data equals a hypothesized variance, we use the χ^2 distribution. The formula is

$$\chi^2 = \frac{(n-1)s^2}{\sigma^2}$$

where

σ^2 = The hypothesized variance

s^2 = The sample variance

n = The sample size

$\chi^2_{\alpha,n-1}$ = Upper critical value for a single-tail test

$\chi^2_{1-\alpha,n-1}$ = Lower critical value for single-tail test

$\chi^2_{\alpha/2,n-1}$ = Upper critical value for a double-tail test

$\chi^2_{1-\alpha/2,n-1}$ = Lower critical value for double-tailed test

Distribution of the chi-square value for a given confidence level for $(n-1)$ degrees of freedom (see Appendix F, Distribution of the Chi-Square)

Example: A process has a known variance (σ^2) of 0.068. A new method of production, suggested to reduce processing time, is being considered. The new method resulted in a variance of 0.087 from 12 samples. We wish to compare the two processes to see if the new production method is the same as the old process method at 10% α risk level.

$H_0 : \sigma_1 = \sigma_2$ (null hypothesis two-tail test)

$H_A : \sigma_1 \neq \sigma_2$ (alternative hypothesis two-tail test)

$\alpha = 0.10$ (two-tail test)

$\chi^2_{\alpha/2,n-1} = 19.675$

$\chi^2_{1-\alpha/2,n-1} = 4.575$

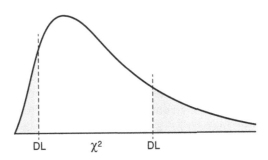

χ^2 **calculated**	χ^2 **critical values**
$\chi^2 = \dfrac{(n-1)s^2}{\sigma^2} = \dfrac{(12-1)0.087}{0.068} = 14.074$	4.575 to 19.675

Since χ^2 calculated falls within the χ^2 critical values at the 0.05 α (alpha) level, there is insufficient evidence to reject the null hypothesis that $H_0 : \sigma_1 = \sigma_2$.

If the requirement were that process 1 is equal to or less than process 2, we would state the null and alternative hypotheses in a different manner, and the result would be a one-sided test with all the risk in the right tail of the distribution.

We would state the hypotheses as:

$H_0 : \sigma_1 \leq \sigma_2$ (null hypothesis one-tail test)

$H_A : \sigma_1 > \sigma_2$ (alternative hypothesis one-tail test)

$\alpha = 0.10$ (one-tail test)

$\chi^2_{\alpha,n-1} = 17.275$

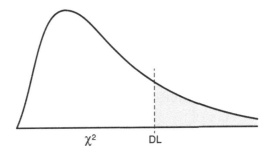

χ^2 calculated	χ^2 critical value

$$\chi^2 = \frac{(n-1)s^2}{s^2} = \frac{(12-1)0.087}{0.068} = 14.074 \quad < \quad 17.275$$

Since χ^2 calculated falls within the χ^2 critical values at the 0.05 α (alpha) level, there is insufficient evidence to reject the null hypothesis that $H_0 : \sigma_1 \leq \sigma_2$.

If the requirement were that process 1 is equal to or greater than process 2, we would state the null and alternative hypotheses in a different manner, and the result would be a one-sided test with all the risk in the left tail of the distribution.

We would state the hypotheses as

$H_0 : \sigma_1 \geq \sigma_2$ (null hypothesis one-tail test)

$H_A : \sigma_1 < \sigma_2$ (alternative hypothesis one-tail test)

$\alpha = 0.10$ (one-tail test)

$\chi^2_{1-\alpha,n-1} = 5.578$

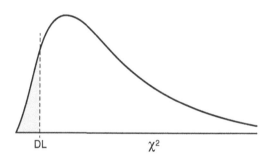

χ^2 calculated	χ^2 critical value

$$\chi^2 = \frac{(n-1)s^2}{\sigma^2} = \frac{(12-1)0.087}{0.068} = 14.074 \quad > \quad 5.578$$

Since χ^2 calculated falls within the χ^2 critical values at the 0.05 α (alpha) level, there is insufficient evidence to reject the null hypothesis that $H_0 : \sigma_1 \geq \sigma_2$.

9.2 TESTING AN OBSERVED VARIANCE AGAINST A HYPOTHESIZED VARIANCE—LARGE SAMPLES

When the sample size is large (> 100), we may use a large sample approximation to test a sample variance versus a hypothesized variance using the following formula:

$$Z = \frac{s - \sigma}{\sigma / \sqrt{2n}}$$

where

σ = Hypothesized standard deviation

s = Standard deviation calculated from sample data

n = Number of samples

Z_α = The normal distribution value for a given confidence level (see Appendix C, Selected Single-Sided Normal Distribution Probability Points)

$Z_{\alpha/2}$ = The normal distribution value for a given confidence level (see Appendix D, Selected Double-Sided Normal Distribution Probability Points)

Example: A process variance has a standard deviation (σ) of 0.261. A new method of production, suggested to reduce processing time, is being considered. The new method resulted in a standard deviation of 0.295 from 100 samples. We wish to compare the two processes to see if the new production method is the same as the old process method at 5% α risk level.

$H_0 : \sigma_1 = \sigma_2$ (null hypothesis two-tail test)

$H_A : \sigma_1 \neq \sigma_2$ (alternative hypothesis two-tail test)

$Z_{\alpha/2} = \pm 1.960$

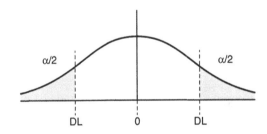

Z calculated	**Z critical**

$$Z = \frac{s-\sigma}{\sigma/\sqrt{2n}} = \frac{0.295-0.261}{0.261/\sqrt{2*100}} = 1.842 \quad < \quad \pm 1.960$$

Since Z calculated is less than Z critical at the 0.05 α (alpha) level, there is insufficient evidence to reject the null hypothesis that $H_0 : \sigma_1 = \sigma_2$.

If the requirement were that process 1 is equal to or less than process 2, we would state the null and alternative hypotheses in a different manner, and the result would be a one-sided test with all the risk in the right tail of the distribution.

We would state the hypotheses as:

$H_0 : \sigma_1 \leq \sigma_2$ (null hypothesis one-tail test)

$H_A : \sigma_1 > \sigma_2$ (alternative hypothesis one-tail test)

$Z_\alpha = 1.645$

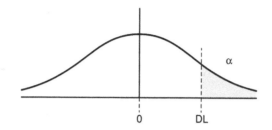

Z calculated		**Z critical**

$$Z = \frac{s - \sigma}{\sigma / \sqrt{2n}} = \frac{0.295 - 0.261}{0.261 / \sqrt{2*100}} = 1.842 \quad > \quad 1.645$$

Since Z calculated is greater than Z critical at the 0.05 α (alpha) level, there is sufficient evidence to reject the null hypothesis $H_0 : \sigma_1 \leq \sigma_2$ in favor of the alternative hypothesis that $H_A : \sigma_1 > \sigma_2$.

If the requirement were that process 1 is equal to or greater than process 2, we would state the null and alternative hypotheses in a different manner, and the result would be a one-sided test with all the risk in the left tail of the distribution.

We would state the hypotheses as

$H_0 : \sigma_1 \geq \sigma_2$ (null hypothesis one-tail test)

$H_A : \sigma_1 < \sigma_2$ (alternative hypothesis one-tail test)

$Z_\alpha = -1.645$

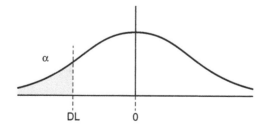

Z calculated	**Z critical**

$$Z = \frac{s - \sigma}{\sigma / \sqrt{2n}} = \frac{0.295 - 0.261}{0.261 / \sqrt{2*100}} = 1.842 \quad > \quad -1.645$$

Since Z calculated is less than Z critical at the 0.05 α (alpha) level, there is insufficient evidence to reject the null hypothesis that $H_0 : \sigma_1 \geq \sigma_2$.

9.3 TESTING FOR A DIFFERENCE BETWEEN TWO OBSERVED VARIANCES USING SAMPLE DATA

When comparing the variances of two populations using sample data, the F-test is appropriate if both populations are normally distributed. While the t-test is less sensitive to small departures from normality, these can significantly affect the risks involved with the F-test. Another assumption of this test is that the samples are independent. The formula for the F-test is

$$F = \frac{s_1^2}{s_2^2}$$

where

$s_1^2 > s_2^2$

$\text{df}_1 = n_1 - 1$

$\text{df}_2 = n_2 - 1$

s_1 = Standard deviation of population 1

s_2 = Standard deviation of population 2

n_1 = Number of samples drawn from population 1

n_2 = Number of samples drawn from population 2

$F\alpha$, $\text{df}_1 = n_1 - 1$, $\text{df}_2 = n_2 - 1$ = values of the F-distribution. (See Appendix G, Percentages of the F-Distribution)

The larger variance is always placed in the numerator. This results in a one-tail test that simplifies both calculation and interpretation.

Example: Two competing machine tool manufacturers presented equipment proposals to a customer. After reviewing both proposals, a test was conducted to determine whether the variances of the product produced by the two machines are equivalent. A sample of 25 from machine 1 produced a variance of 0.1211, and 30 units from machine 2 resulted in a variance of 0.0707. At the 95% confidence level, test the hypothesis that the variances are equal.

$H_0 : \sigma_1 = \sigma_2$ (null hypothesis two-tail test)

$H_A : \sigma_1 \neq \sigma_2$ (alternative hypothesis two-tail test)

$F\alpha$, $df_1 = n_1 - 1$, $df_2 = n_2 - 1 = 1.90$

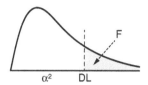

F **calculated**	*F* **critical**

$$F = \frac{s_1^2}{s_2^2} = \frac{0.1211}{0.0701} = 1.728 \quad < \quad 1.90$$

Since F calculated is less than F critical at the 0.05 α (alpha) level, there is insufficient evidence to reject the null hypothesis that $H_0 : \sigma_1 = \sigma_2$.

9.4 TESTING FOR A DIFFERENCE BETWEEN TWO OBSERVED VARIANCES USING LARGE SAMPLES

When the sample sizes are large (> 100), we may use a large sample approximation of the F-test. The formula for this approximation is:

$$Z = \frac{s_1 - s_2}{\sqrt{\dfrac{s_1^2}{2(n_2 - 1)} + \dfrac{s_2^2}{2(n_1 - 1)}}}$$

where

s_1 = Standard deviation of population 1

s_2 = Standard deviation of population 2

n_1 = Number of samples drawn from population 1

n_2 = Number of samples drawn from population 2

Z_α = The normal distribution value for a given confidence level (see Appendix C, Selected Single-Sided Normal Distribution Probability Points)

$Z_{\alpha/2}$ = The normal distribution value for a given confidence level (see Appendix D, Selected Double-Sided Normal Distribution Probability Points)

Note the switching of the sample sizes in the denominator.

Example: Two competing machine tool manufacturers presented equipment proposals to a customer. After reviewing both proposals, a test was conducted to determine whether the variances of the product produced by the two machines are equivalent. A sample of 150 from machine 1 produced a variance of 0.361, and 125 units from machine 2 resulted in a variance of 0.265. At the 95% confidence level, test the hypothesis that the variances are equal.

$H_0 : \sigma_1 = \sigma_2$ (null hypothesis two-tail test)

$H_A : \sigma_1 \neq \sigma_2$ (alternative hypothesis two-tail test)

$Z_{\alpha/2} = 1.960$

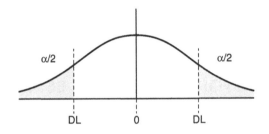

Z calculated	**Z critical**

$$Z = \frac{s_1 - s_2}{\sqrt{\dfrac{s_1^2}{2(n_2 - 1)} + \dfrac{s_2^2}{2(n_1 - 1)}}} = \frac{0.361 - 0.265}{\sqrt{\dfrac{0.361^2}{2(125 - 1)} + \dfrac{0.265^2}{2(150 - 1)}}} = 3.480 \quad > \quad \pm 1.960$$

Since Z calculated is greater than Z critical at the 0.05 α (alpha) level, there is sufficient evidence to reject the null hypothesis $H_0 : \sigma_1 = \sigma_2$ in favor of the alternative hypothesis that $H_A : \sigma_1 \neq \sigma_2$.

If the requirement were that machine 1 be less than or equal to machine 2, we would state the null and alternative hypotheses in a different manner, and the result would be a one-sided test with all the risk in the right tail of the distribution.

We would state the hypotheses as

$H_0 : \sigma_1 \leq \sigma_2$ (null hypothesis one-tail test)

$H_A : \sigma_1 > \sigma_2$ (alternative hypothesis one-tail test)

$Z_\alpha = 1.645$

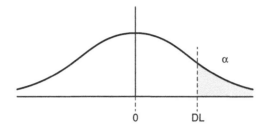

Z calculated		Z critical

$$Z = \frac{s_1 - s_2}{\sqrt{\dfrac{s_1^2}{2(n_2 - 1)} + \dfrac{s_2^2}{2(n_1 - 1)}}} = \frac{0.361 - 0.265}{\sqrt{\dfrac{0.361^2}{2(125 - 1)} + \dfrac{0.265^2}{2(150 - 1)}}} = 3.480 \quad > \quad 1.645$$

Since *Z* calculated is greater than *Z* critical at the 0.05 α (alpha) level, there is sufficient evidence to reject the null hypothesis $H_0 : \sigma_1 \leq \sigma_2$ in favor of the alternative hypothesis that $H_A : \sigma_1 > \sigma_2$.

If the requirement were that machine 1 be greater than or equal to machine 2, we would state the null and alternative hypotheses in a different manner, and the result would be a one-sided test with all the risk in the left tail of the distribution.

We would state the hypotheses as

$H_0 : \sigma_1 \geq \sigma_2$ (null hypothesis one-tail test)

$H_A : \sigma_1 < \sigma_2$ (alternative hypothesis one-tail test)

$Z_\alpha = -1.645$

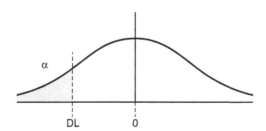

Z calculated		Z critical

$$Z = \frac{s_1 - s_2}{\sqrt{\dfrac{s_1^2}{2(n_2 - 1)} + \dfrac{s_2^2}{2(n_1 - 1)}}} = \frac{0.361 - 0.265}{\sqrt{\dfrac{0.361^2}{2(125 - 1)} + \dfrac{0.265^2}{2(150 - 1)}}} = 3.480 \quad > \quad -1.645$$

Since *Z* calculated is less than *Z* critical at the 0.05 α (alpha) level, there is insufficient evidence to reject the null hypothesis that $H_0 : \sigma_1 \geq \sigma_2$.

10
Discrete Probability Distributions

10.1 BINOMIAL DISTRIBUTION

Using the *binomial distribution* is appropriate when there is a series of n trials, and each trial has only two possible outcomes: the occurrence of an event (x) or the nonoccurrence of the event. An example of an occurrence might be a nonconforming item.

Some assumptions when using the binomial distribution are:

1. The probability of an occurrence (x) and a nonoccurrence remain constant from trial to trial.

2. The trials are independent.

The first condition is met when the lot size is infinite or sampling is with replacement. Juran (1988) finds that this can be approximated in finite populations when the population size is at least 10 times the sample size.

The second condition states that the drawing of a sample and its outcome have no effect on the outcome of any other sample, nor is the sample affected by any previous sample.

When working with the binomial distribution, the probability of an occurrence is designated as p. The probability of a nonoccurrence is $(1 - p)$.

p = Probability of occurrence

$1 - p$ = Probability of nonoccurrence

We can generalize the binomial expansion into a general expression that allows the calculation of exactly x occurrences in a series on n trials:

$$P(x) = C_X^n p^x (1 - p)^{n-x}$$

where

$$C_X^n = \frac{n!}{x!(n-x)!}$$

p = Probability of occurrence

$1 - p$ = Probability of nonoccurrence

n = Sample size

x = Number of occurrences in n trials

Example: A process has produced nonconformances at a rate of 3% (p = .03). If a random sample of 15 items is drawn from the process, what is the probability of obtaining exactly three nonconforming items?

$$P(x) = C_x^n p^x (1-p)^{n-x}$$

$$P(3) = C_3^{15} (0.03)^3 (0.97)^{12}$$

$$P(3) = 0.0085$$

The probability of obtaining exactly three occurrences in a sample of 15 is 0.0085 or 0.85%.

This problem can also be solved using a binomial distribution table (see Appendix L). To solve this problem, we look at the values n = 15, 0.03%, and r = 2, which is 0.9992. From this we must subtract the r = 2 value of .9906 (because we are asking for exactly three nonconforming items):

$$0.9992 - 0.9906 = .0086 \text{ or } .86\%$$

(Please note that the difference between the calculated value and the table value is due to rounding.)

At times it is necessary to calculate the probability of x or fewer occurrences in a series of n trials. In this case the probability of x or less is the probability of 0 occurrences in n trials, plus the probability of 1 occurrence in n trials, plus the probability of 2 occurrences in n trials, plus all the way to x occurrences in n trials:

$$P(\leq x) = \sum C_x^n p^x (1-p)^{n-x}$$

where

p = Probability of occurrence

$1 - p$ = Probability of nonoccurrence

n = Sample size

x = Number of occurrences in n trials

Example: A process has produced nonconformances at a rate of 3% (p = 0.03). If a random sample of 15 items is drawn from the process, what is the probability of obtaining two or fewer nonconforming items?

$$P(0) = C_X^n p^x (1-p)^{n-x} = C_0^{15}(0.03)^0 (0.97)^{15} = 0.6333$$

$$P(1) = C_X^n p^x (1-p)^{n-x} = C_1^{15}(0.03)^1 (0.97)^{14} = 0.2938$$

$$P(2) = C_X^n p^x (1-p)^{n-x} = C_2^{15}(0.03)^2 (0.97)^{13} = 0.0636$$

$$\sum = 0.9907$$

The probability of two or fewer occurrences in a sample of 15 is 0.9907 or 99.07%

This problem can also be solved using a binomial distribution table (see Appendix L). To solve this problem we look at the values $n = 15$, 0.03%, and $r = 2$, which is 0.9906 or 99.06%. (Please note that the difference between the calculated value and the table value is due to rounding.)

While performing calculations using the binomial distribution, the sum of all possible probabilities of occurrences sums to 1. Therefore, the probability of three or more occurrences is $1 - 0.9907 = 0.0093$ or 0.93%.

10.2 POISSON DISTRIBUTION

The *Poisson distribution* determines the probability of x occurrences over a given area of opportunity. This could be the number of occurrences over a specified time interval, number of nonconformances per assembly, number of customers per day, and so on.

Assumptions that must be met when using the Poisson probability distribution are:

1. The potential number of occurrences must be very large compared to the expected number of occurrences.

2. The occurrences are independent.

Condition 1 is commonly met in practice with nonconformances per assembly, such as the number of nonconformances per 100 vehicles manufactured.

Condition 2 states that the drawing of a sample and its outcome have no effect on the outcome of any other sample, nor is the sample affected by any previous sample.

The Poisson distribution can be generalized into an expression allowing calculation of exactly x occurrences over a given opportunity. This expression is

$$P(x) = \frac{e^{-\mu}\mu^x}{x!}$$

where

$e = 2.71828...$, the base of natural logarithms

μ = The expected number of occurrences

x = Number of occurrences during this trial

Example: A coil of wire has an average of seven nonconformances per 10,000 feet. What is the probability of nine nonconformances in a randomly selected 10,000-foot coil?

$$P(x) = \frac{e^{-\mu}\mu^{x}}{x!}$$

$$P(9) = \frac{e^{-7}(7^{9})}{9!}$$

$$P(9) = 0.101$$

The probability of nine occurrences when the expected value is seven nonconformances is 0.101 or 10.1%.

This problem can also be solved using a Poisson distribution table (see Appendix M). To solve this problem, we look at the $np = 7$ row and the "9" column, which is 0.830. From this we must subtract the "8" column value of 0.729 (because we are asking for exactly nine nonconformances):

$$0.830 - 0.729 = 0.101 \text{ or } 10.1\%$$

At times it is necessary to calculate the probability of x or fewer occurrences. In this case the probability of x or less is the probability of 0 occurrences, plus the probability of 1 occurrence, plus the probability of 2 occurrences, plus all the way to x occurrences:

$$P(\le x) = \sum \frac{e^{-\mu}\mu^{x}}{x!}$$

where

e = 2.71828..., the base of natural logarithms

μ = The expected number of occurrences

x = Number of occurrences during this trial

Example: A coil of wire has an average of seven nonconformances per 10,000 feet. What is the probability of two or fewer nonconformances in a randomly selected 10,000-foot coil?

$$P(0) = \frac{e^{-\mu}\mu^{x}}{x!} = \frac{e^{-7}(7^{0})}{0!} = 0.0009$$

$$P(1) = \frac{e^{-\mu}\mu^{x}}{x!} = \frac{e^{-7}(7^{1})}{1!} = 0.0064$$

$$P(2) = \frac{e^{-\mu}\mu^{x}}{x!} = \frac{e^{-7}(7^{2})}{2!} = 0.0223$$

$$\sum = 0.0296$$

The probability of two or fewer occurrences when the expected number of nonconformances equals seven is 0.0296 or 2.96%.

This problem can also be solved using a Poisson distribution table (see Appendix M). To solve this problem we look at the $np = 7$ row and the "2" column, which is 0.030 or 3.0%. (Please note that the difference between the calculated value and the table value is due to rounding.)

While performing calculations using the Poisson distribution, the sum of all possible probabilities of occurrences sums to 1. Therefore, the probability of two or more occurrences is $1 - 0.030 = 0.97$ or 97%.

10.3 HYPERGEOMETRIC DISTRIBUTION

When the population is finite and we draw a series of samples from the population, the probability of occurrence varies from trial to trial, so using the binomial distribution is no longer appropriate.

We use the *hypergeometric distribution* when there is a series of n trials from a finite population, when each trial has only two possible outcomes, and we sample without replacement.

We calculate the probability of x occurrences in a sample size (n) from a finite population size (N) containing M nonconformances as

$$P(x) = \frac{C_{n-x}^{N-M} C_x^M}{C_n^N}$$

where

M = Population nonconformances

N = Population size

n = Sample size

x = Number of occurrences in a sample

Example: If a lot size of 25 units contains two nonconforming items, what is the probability of finding one nonconformance in a sample of five units?

$$P(1) = \frac{C_{n-x}^{N-M} C_x^M}{C_n^N}$$

$$P(1) = \frac{C_{5-1}^{25-2} C_1^2}{C_5^{25}}$$

$$P(1) = 0.3333$$

The probability of finding one nonconformance in a sample of five from a lot size of 25 containing two nonconformances is 0.3333 or 33.33%.

At times it is necessary to calculate the probability of *x* or fewer occurrences. In this case the probability of *x* or less is the probability of 0 occurrences, plus the probability of 1 occurrence, plus the probability of 2 occurrences, plus all the way to *x* occurrences:

$$P(\le x) = \sum \frac{C_{n-x}^{N-M} C_x^M}{C_n^N}$$

where

M = Population nonconformances

N = Population size

n = Sample size

x = Number of occurrences in a sample

Example: If a lot size of 25 units contains two nonconforming items, what is the probability of finding one or fewer nonconformances in a sample of five units?

$$P(0) = \frac{C_{n-x}^{N-M} C_x^M}{C_n^N} = \frac{C_{5-0}^{25-2} C_0^2}{C_5^{25}} = 0.6333$$

$$P(1) = \frac{C_{n-x}^{N-M} C_x^M}{C_n^N} = \frac{C_{5-1}^{25-2} C_1^2}{C_5^{25}} = 0.3333$$

$$\sum = 0.9666$$

The probability of one or fewer nonconformances in the sample of five is 0.9666 or 96.66%.

10.4 GEOMETRIC DISTRIBUTION

In many cases the *geometric distribution* is similar to the binomial distribution, as noted by Burr (1974), in that there are *n* independent trials having two possible outcomes, and the probability of occurrence for each outcome is constant from trial to trial.

The departure from the binomial is that the probability of the first occurrence is of interest (*x* is fixed) and the value of *n* varies.

The probability that the first occurrence occurs on the *x*th trial is

$$P(x; p) = (1 - p)^{x-1} p$$

where

p = Probability of occurrence

x = The number of trials

Example: If the probability of a nonconforming item is 0.02, what is the probability that the first failure will occur on the hundredth trial?

$$P(x; p) = (1 - p)^{x-1} p$$

$$P(100; 0.02) = (1 - 0.02)^{10-1} 0.02$$

$$P(100; 0.02) = 0.003$$

The probability that the first failure occurs on the hundredth trial is 0.003 or .3%.

If we desire to calculate the expected number of samples required to the first failure we use the following calculation:

$$E(x) = \frac{1}{p}$$

where

p = Probability of occurrence

If the probability of a nonconforming item is 0.02, what is the expected number of samples required to the first failure?

$$E(x) = \frac{1}{p}$$

$$E(x) = \frac{1}{0.02}$$

$$E(x) = 50$$

The expected number of trials (samples) to the first nonconforming item is 50.

10.5 NEGATIVE BINOMIAL DISTRIBUTION

The negative binomial distribution has application in determining the probability that the xth occurrence occurs on the nth trial. The probability of an occurrence, as in the binomial distribution, is designated as p.

The solution to this is equivalent to calculating the probability of $x - 1$ occurrences in $n - 1$ trials and the xth occurrences at the nth trial:

$$P(x,n) = C_{x-1}^{n-1} p^x (1 - p)^{n-x}$$

where

p = Probability of occurrence

n = Trial

x = Number of occurrences

Example: The probability of occurrence on any trial is 0.03. What is the probability that the second occurrence occurs on the fifth trial?

$$P(x,n) = C_{x-1}^{n-1} p^x (1-p)^{n-x}$$

$$P(2,5) = C_{2-1}^{5-1} p^2 (1-0.03)^{5-2}$$

$$P(2,5) = 0.0032$$

The probability that the second occurrence occurs on the fifth trial is 0.0032 or .32%.

11

Control Charts

*C*ontrol charts are decision-making tools that provide information for timely decisions concerning recently produced products. Control charts are also problem-solving tools that help locate and investigate the causes of poor or marginal quality.

Control charts contain a centerline—usually the mathematical average of the samples plotted—upper and lower statistical control limits that define the constraints of common cause variation, and performance data plotted over time.

11.1 CONTROL CHART TYPES AND SELECTION

There are two general classifications of control charts: variables charts and attributes charts (see Table 11.1). *Variables* are things that can be measured: length, temperature, pressure, weight, and so on. *Attributes* are things that are counted: dents, scratches, defects, days, cycles, yes/no decisions, and so on.

Table 11.1 Variables and attributes control charts selection.

Variables control charts

Type	Distribution	Sample	Application
$\overline{X}$ and R	Normal	$2 \leq 10$	Measurement subgroups
$\overline{X}$ and s	Normal	> 10	Measurement subgroups

Attributes control charts

Type	Distribution	Sample	Application
c	Poisson	Constant	Count number of defects per item
u	Poisson	Varies	Count number of defects per item
np	Binomial	Constant	Count of defective items
p	Binomial	Varies	Count of defective items
g	Binomial	Individual	Interval between rare events

Variables or attributes control charts

Type	Distribution	Sample	Application
X and mR	Normal	1	Individual counts or measurements

11.2 CONTROL CHART INTERPRETATION

A process is said to be *in control* when the control chart does not indicate any out-of-control condition and contains only common causes of variation. If the common cause variation is small, then a control chart can be used to monitor the process. See Figure 11.1 for a representation of stable (in control) and unstable (out of control) processes. If the common cause variation is too large, the process will need to be modified.

When a control chart indicates an out-of-control condition (a point outside the control limits or matching one or more of the criteria below), the assignable causes of variation must be identified and eliminated.

Improper control chart interpretation can lead to several problems, including blaming people for problems they can not control, spending time and money looking for problems that do not exist, spending time and money on process adjustments or new equipment that are not necessary, taking action where no action is warranted, and asking for worker-related improvements where process or equipment improvements need to be made first.

The following rules should be used to properly interpret control charts:

- Rule 1—One point beyond the 3 σ control limit

- Rule 2—Eight or more points on one side of the centerline without crossing

- Rule 3—Four out of five points in zone B or beyond

- Rule 4—Six points or more in a row steadily increasing or decreasing

- Rule 5—Two out of three points in zone A

- Rule 6—14 points in a row alternating up and down

- Rule 7—Any noticeable/predictable pattern, cycle, or trend

Please note that depending on the source, these rules can vary. See Figure 11.2 for control chart interpretation rules.

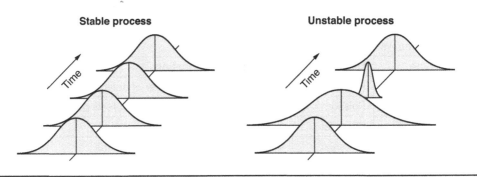

Figure 11.1   Stable and unstable variation.

Source: Adapted from Stephen A. Wise and Douglas C. Fair, *Innovative Control Charting: Practical SPC Solutions for Today's Manufacturing Environment* (Milwaukee: ASQ Quality Press, 1997), Figures 3.3 and 3.5. Used with permission.

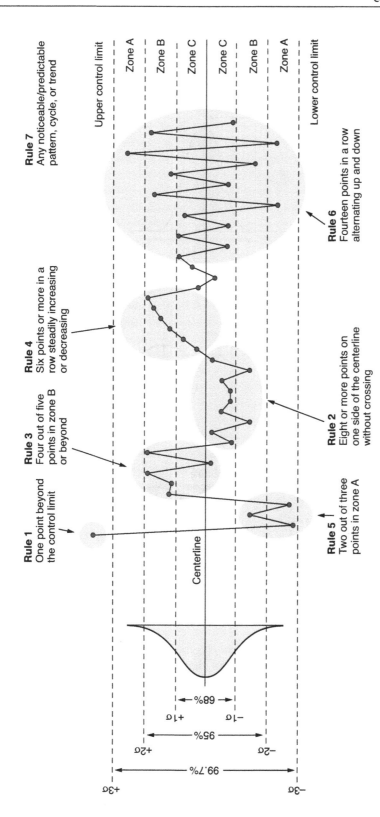

Figure 11.2 Control chart interpretation rules.

Source: Adapted from Stephen A. Wise and Douglas C. Fair, *Innovative Control Charting: Practical SPC Solutions for Today's Manufacturing Environment* (Milwaukee: ASQ Quality Press, 1997), Figures 3.3 and 3.5. Used with permission.

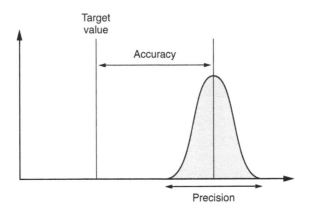

Figure 11.3   Control chart accuracy and precision.

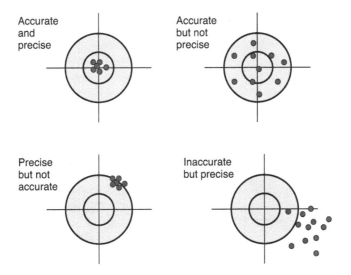

Figure 11.4   Accuracy versus precision.

The target value (which is hopefully the control chart centerline) is closely related to process *accuracy*. The range chart is closely associated with process *precision* (spread or dispersion) (see Figures 11.3 and 11.4).

11.3 $\bar{X}$ AND R CONTROL CHARTS

$\bar{X}$ and R control charts assume a normal distribution and are usually used with a subgroup size of less than 10 (typically 3–5). A minimum of 25 subgroups is necessary to construct the chart.

$\bar{X}$ **chart**	R **chart**
$\bar{X} = \dfrac{\Sigma X}{n}$ (Subgroup)	$R = X_n - X_i$ (Subgroup)
$\bar{\bar{X}} = \dfrac{\Sigma \bar{X}}{k}$ (Centerline)	$\bar{R} = \dfrac{\Sigma R}{k}$ (Centerline)
$\bar{X}\,\text{UCL} = \bar{\bar{X}} + A_2 * \bar{R}$	$\bar{R}\,\text{UCL} = D_4 * \bar{R}$
$\bar{X}\,\text{LCL} = \bar{\bar{X}} - A_2 * \bar{R}$	$\bar{R}\,\text{LCL} = D_3 * \bar{R}$

$$s = \frac{\bar{R}}{d_2} \quad \text{(Estimate of sigma)}$$

where

n = Subgroup size

k = The number of subgroups

(See Appendix O, Control Chart Constants.)

Example: A process has an $\bar{X}$ of 5.496 and an $\bar{R}$ of 0.065. Twenty-five subgroups of three samples are taken (Table 11.2). Calculate the centerlines and upper and lower control limits, and construct an $\bar{X}$ and R chart (Figure 11.5).

$$\bar{\bar{X}} = \frac{\Sigma \bar{X}}{k} = \frac{137.390}{25} = 5.496$$

$$\bar{R} = \frac{\Sigma R}{k} = \frac{1.620}{25} = 0.065$$

$$\bar{X}\,\text{UCL} = \bar{\bar{X}} + A_2 * \bar{R} = 5.496 + 1.023 * 0.065 = 5.535$$

$$\bar{X}\,\text{UCL} = \bar{\bar{X}} + A_2 * \bar{R} = 5.496 - 1.023 * 0.065 = 5.430$$

$$\bar{R}\,\text{UCL} = D_4 * \bar{R} = 2.574 * 0.065 = .167$$

$$\bar{R}\,\text{LCL} = D_3 * \bar{R} = 0 * 0.065 = 0$$

Table 11.2 Data for $\bar{X}$ and R chart.

Subgroup	$\bar{X}$	R
1	5.447	0.080
2	5.487	0.100
3	5.450	0.060
4	5.450	0.060
5	5.477	0.100
6	5.510	0.020
7	5.510	0.030
8	5.463	0.080
9	5.553	0.040
10	5.510	0.110
11	5.627	0.030
12	5.610	0.110
13	5.507	0.090
14	5.497	0.070
15	5.540	0.080
16	5.413	0.060
17	5.490	0.020
18	5.490	0.060
19	5.467	0.060
20	5.467	0.060
21	5.500	0.070
22	5.477	0.070
23	5.490	0.030
24	5.470	0.080
25	5.490	0.050
Sum	**137.390**	**1.620**
Average	**5.496**	**0.065**

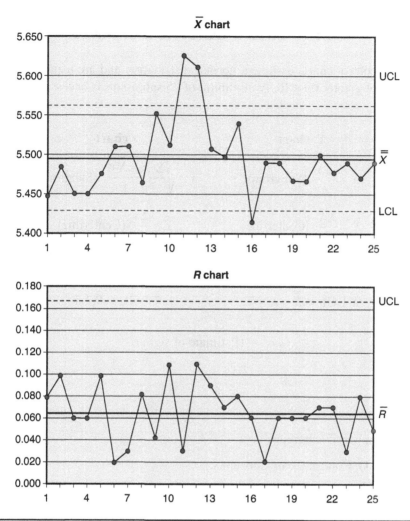

Figure 11.5 $\overline{X}$ and R chart example.

11.4 $\bar{X}$ AND s CONTROL CHARTS

The $\bar{X}$ and s control charts assume a normal distribution and are usually used with a subgroup size of greater than 10. A minimum of 25 subgroups is necessary to construct the chart.

$\bar{X}$ chart	s chart

$$\bar{X} = \frac{\Sigma X}{n} \text{ (Subgroup)} \qquad s = \sqrt{\frac{\Sigma(X - \bar{X})^2}{n-1}} \text{ (Subgroup)}$$

$$\bar{\bar{X}} = \frac{\Sigma \bar{X}}{k} \text{ (Centerline)} \qquad \bar{s} = \frac{\Sigma s}{k} \text{ (Centerline)}$$

$$\bar{X}\,\text{UCL} = \bar{\bar{X}} + A_3 * \bar{s} \qquad \bar{s}\,\text{UCL} = B_4 * \bar{s}$$

$$\bar{X}\,\text{LCL} = \bar{\bar{X}} - A_3 * \bar{s} \qquad \bar{s}\,\text{LCL} = B_3 * \bar{s}$$

$$s = \frac{\bar{s}}{c_4} \text{ (Estimate of sigma)}$$

where

n = Subgroup size

k = The number of subgroups

(See Appendix O, Control Chart Constants.)

Example: A process has an $\bar{X}$ of 5.498 and an $\bar{s}$ of 0.046. Twenty-five subgroups of 11 samples are taken (Table 11.3). Calculate the centerlines and the upper and lower control limits, and construct and $\bar{X}$ and s chart (Figure 11.6).

$$\bar{\bar{X}} = \frac{\Sigma \bar{X}}{k} = \frac{137.488}{25} = 5.498$$

$$\bar{s} = \frac{\Sigma s}{k} = \frac{1.155}{25} = 0.046$$

$$\bar{X}\,\text{UCL} = \bar{\bar{X}} + A_3 * \bar{s} = 5.498 + 0.927 * 0.046 = 5.541$$

$$\bar{X}\,\text{LCL} = \bar{\bar{X}} - A_3 * \bar{s} = 5.498 - 0.927 * 0.046 = 5.455$$

$$\bar{s}\,\text{UCL} = B_4 * \bar{s} = 1.679 * 0.046 = 0.077$$

$$\bar{s}\,\text{LCL} = B_3 * \bar{s} = 0.321 * 0.046 = 0.015$$

Table 11.3 Data for $\bar{X}$ and s chart.

Subgroup	$\bar{X}$	s
1	5.449	0.032
2	5.487	0.039
3	5.456	0.037
4	5.612	0.121
5	5.479	0.033
6	5.535	0.106
7	5.516	0.016
8	5.465	0.045
9	5.551	0.036
10	5.520	0.031
11	5.489	0.030
12	5.474	0.026
13	5.481	0.035
14	5.480	0.031
15	5.537	0.030
16	5.439	0.047
17	5.505	0.098
18	5.485	0.080
19	5.475	0.071
20	5.590	0.057
21	5.487	0.040
22	5.489	0.039
23	5.492	0.021
24	5.475	0.025
25	5.479	0.029
Sum	**137.448**	**1.155**
Average	**5.498**	**0.046**

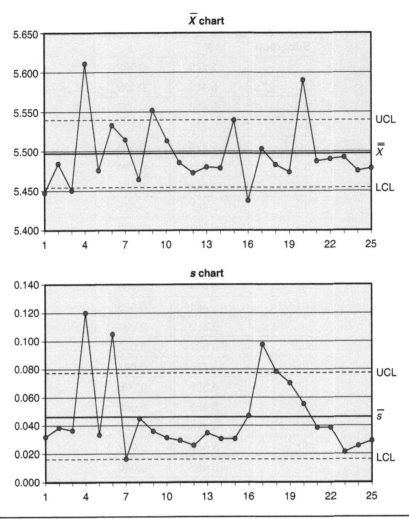

Figure 11.6 $\overline{X}$ and *s* chart example.

11.5 *c*-CHARTS

c-charts assume a Poisson distribution and are usually used with a constant sample size, counting the number of defects per item. A minimum of 25 subgroups is necessary to construct the chart.

$$c = \text{(Subgroup count)}$$

$$\bar{c} = \frac{\Sigma c}{k} \text{ (Centerline)}$$

$$\text{UCL} = \bar{c} + 3 * \sqrt{\bar{c}}$$

$$\text{LCL} = \bar{c} - 3 * \sqrt{\bar{c}}$$

(A calculated LCL of less than zero reverts to zero.)

where

k = The number of subgroups

Example: A process is evaluated using a constant sample size of 48 items. When the items are inspected, a count of the number of defects is recorded (Table 11.4). Twenty-five subgroups of 48 samples are taken. Calculate the centerline and the upper and lower control limits, and construct a *c*-chart (Figure 11.7).

$$\bar{c} = \frac{\Sigma c}{k} = \frac{136}{25} = 5.440$$

$$\text{UCL} = \bar{c} + 3 * \sqrt{\bar{c}} = 5.440 + 3 * \sqrt{5.440} = 12.437$$

$$\text{LCL} = \bar{c} - 3 * \sqrt{\bar{c}} = 5.440 - 3 * \sqrt{5.440} = -1.557 \text{ reverts to } 0$$

Table 11.4   Data for c-chart.

Subgroup	c
1	6
2	7
3	5
4	2
5	11
6	7
7	7
8	0
9	1
10	4
11	15
12	5
13	0
14	6
15	3
16	7
17	6
18	8
19	1
20	7
21	1
22	7
23	11
24	3
25	6
Sum	**136**
Average	**5.440**

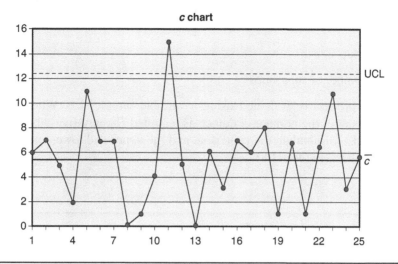

Figure 11.7 c-chart example.

11.6 *u*-CHARTS

u-charts assume a Poisson distribution and are usually used with a variable sample size (although a constant sample size may be used), counting the number of defects per item. A minimum of 25 subgroups is necessary to construct the chart.

$$u = \frac{c}{n} \text{ (Subgroup)}$$

$$\bar{u} = \frac{\Sigma c}{\Sigma n} \text{ (Centerline)}$$

(Floating control limits)   (Static control limits)

$$\text{UCL} = \bar{u} + 3 * \sqrt{\frac{\bar{u}}{n}}$$ $$\text{UCL} = \bar{u} + 3 * \sqrt{\frac{\bar{u}}{\Sigma n / k}}$$

$$\text{LCL} = \bar{u} - 3 * \sqrt{\frac{\bar{u}}{n}}$$ $$\text{LCL} = \bar{u} - 3 * \sqrt{\frac{\bar{u}}{\Sigma n / k}}$$

(A calculated LCL of less than zero reverts to zero.)

$$s \cong \sqrt{\frac{\bar{u}}{n}}$$ $$s \cong \sqrt{\frac{\bar{u}}{\Sigma n / k}}$$

(Estimate of sigma for subgroup)   (Estimate of sigma for chart)

where

n = Subgroup size

k = The number of subgroups

Example: A process is evaluated using a variable sample size. When the items are inspected, a count of the number of defects is recorded. Twenty-five subgroups are evaluated (Table 11.5). Calculate the centerline and the upper and lower control limits, and construct a u-chart (Figure 11.8).

Table 11.5 Data for u-chart.

Subgroup	n	c	u	UCL	LCL
1	113	77	0.681	1.232	−1.977
2	85	72	0.847	1.274	−1.118
3	99	125	1.263	1.251	−0.737
4	118	121	1.025	1.226	−0.511
5	111	80	0.721	1.235	−0.356
6	59	79	1.339	1.338	−0.241
7	123	117	0.951	1.221	−0.153
8	101	80	0.792	1.248	−0.081
9	105	128	1.219	1.243	−0.022
10	118	111	0.941	1.226	0.029
11	91	66	0.725	1.264	0.072
12	74	103	1.392	1.297	0.109
13	98	99	1.010	1.253	0.143
14	106	110	1.038	1.241	0.172
15	116	75	0.647	1.229	0.199
16	88	123	1.398	1.269	0.223
17	104	43	0.413	1.244	0.245
18	45	77	1.711	1.394	0.265
19	100	91	0.910	1.250	0.283
20	95	114	1.200	1.257	0.300
21	103	102	0.990	1.245	0.316
22	118	76	0.644	1.226	0.331
23	102	71	0.696	1.247	0.345
24	60	90	1.500	1.335	0.357
25	92	88	0.957	1.262	0.370
Sum	**2424**	**2318**			

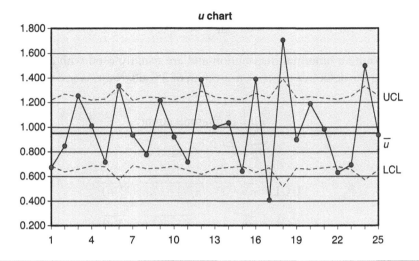

Figure 11.8   *u*-chart example.

$$\bar{u} = \frac{\Sigma c}{\Sigma n} = \frac{2318}{2424} = 0.956$$

$$\text{UCL} = \bar{u} + 3 * \sqrt{\frac{\bar{u}}{n}}$$

$$\text{LCL} = \bar{u} - 3 * \sqrt{\frac{\bar{u}}{n}}$$

The control limits will vary for each subgroup because the sample size n varies from group to group.

11.7 *np*-CHARTS

np-charts assume a binomial distribution and are usually used with a constant sample size, counting the defective items. A minimum of 25 subgroups is necessary to construct the chart.

$$np = (\text{Subgroup count})$$

$$\overline{np} = \frac{\Sigma np}{k} \quad (\text{Centerline})$$

$$\text{UCL} = \overline{np} + 3 * \sqrt{n\overline{p}(1 - (\Sigma np / \Sigma n))}$$

$$\text{LCL} = \overline{np} - 3 * \sqrt{n\overline{p}(1 - (\Sigma np / \Sigma n))}$$

(A calculated LCL of less than zero reverts to zero.)

$$s \cong \sqrt{n\overline{p}(1 - (\Sigma np / \Sigma n))} \quad (\text{Estimate of sigma})$$

where

n = Subgroup size

k = The number of subgroups

Example: A process is evaluated using a constant sample size of 200 items. When the items are inspected, a count of the number of defective items is recorded. Twenty-five subgroups of 200 samples are taken (Table 11.6). Calculate the centerline and the upper and lower control limits, and construct an *np*-chart (Figure 11.9).

$$\overline{np} = \frac{\Sigma np}{k} = \frac{100}{25} = 4.000$$

$$\text{UCL} = \overline{np} + 3 * \sqrt{n\overline{p}(1 - (\Sigma np / \Sigma n))} = 4.000 + 3 * \sqrt{4.000(1 - (100 / 5000))} = 9.940$$

$$\text{LCL} = \overline{np} - 3 * \sqrt{n\overline{p}(1 - (\Sigma np / \Sigma n))} = 4.000 - 3 * \sqrt{4.000(1 - (100 / 5000))} = -1.940,$$

which reverts to zero.

Table 11.6   Data for *np*-chart.

Subgroup	*n*	*np*
1	200	2
2	200	5
3	200	3
4	200	5
5	200	11
6	200	1
7	200	5
8	200	6
9	200	2
10	200	0
11	200	5
12	200	8
13	200	4
14	200	2
15	200	6
16	200	3
17	200	4
18	200	6
19	200	4
20	200	1
21	200	1
22	200	2
23	200	5
24	200	0
25	200	9
Sum	**5000**	**100**

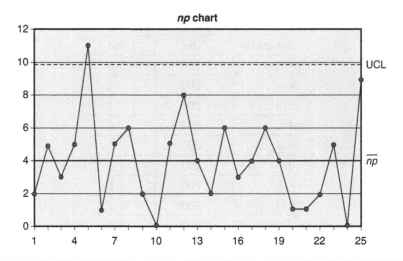

Figure 11.9   *np*-chart example.

11.8 *p*-CHARTS

p-charts assume a binomial distribution and are usually used with a variable sample size (although a constant sample size may be used), counting the defective items. A minimum of 25 subgroups is necessary to construct the chart.

$$p = \frac{np}{n} \ \text{(Subgroup)}$$

$$\bar{p} = \frac{\Sigma np}{\Sigma n} \ \text{(Centerline)}$$

(Floating control limits)	(Static control limits)
$\text{UCL} = \bar{p} + 3 * \sqrt{(\bar{p}(1-\bar{p}))/n}$	$\text{UCL} = \bar{p} + 3 * \sqrt{(\bar{p}(1-\bar{p}))/(\Sigma n/k)}$
$\text{LCL} = \bar{p} - 3 * \sqrt{(\bar{p}(1-\bar{p}))/n}$	$\text{LCL} = \bar{p} - 3 * \sqrt{(\bar{p}(1-\bar{p}))/(\Sigma n/k)}$

(A calculated LCL of less than zero reverts to zero.)

$$s \cong \sqrt{(\bar{p}(1-\bar{p}))/n} \qquad\qquad s \cong \sqrt{(\bar{p}(1-\bar{p}))/(\Sigma n/k)}$$

(Estimate of sigma for subgroup)   (Estimate of sigma for chart)

where

n = Subgroup size

k = The number of subgroups

Example: A process is evaluated using a variable sample size. When the items are inspected, a count of the number of defective items is recorded. Twenty-five subgroups are evaluated (Table 11.7). Calculate the centerline and the upper and lower control limits, and construct a p-chart (Figure 11.10).

Table 11.7 Data for p-chart.

Subgroup	*n*	*np*	*p*	UCL	LCL
1	2108	254	0.120		
2	1175	117	0.100	0.142	0.203
3	1658	248	0.150	0.138	0.175
4	2173	239	0.110	0.135	0.176
5	1720	238	0.138	0.138	0.176
6	1891	250	0.132	0.136	0.175
7	1664	145	0.087	0.138	0.194
8	1685	127	0.075	0.138	0.199
9	1967	327	0.166	0.136	0.167
10	976	95	0.097	0.145	0.212
11	2012	201	0.100	0.136	0.182
12	1187	124	0.104	0.142	0.200
13	1784	173	0.097	0.137	0.187
14	1390	132	0.095	0.140	0.198
15	2075	229	0.110	0.135	0.178
16	2077	236	0.114	0.135	0.177
17	1232	126	0.102	0.142	0.200
18	1914	228	0.119	0.136	0.178
19	2401	315	0.131	0.134	0.168
20	1975	255	0.129	0.136	0.174
21	1365	132	0.097	0.140	0.198
22	1505	117	0.078	0.139	0.203
23	1725	129	0.075	0.137	0.199
24	2105	315	0.150	0.135	0.168
25	1855	242	0.130	0.137	0.176
Sum	**43619**	**4994**			

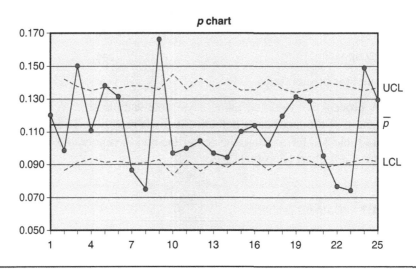

Figure 11.10   *p*-chart example.

$$\bar{p} = \frac{\Sigma np}{\Sigma n} = \frac{4994}{43619} = 0.114$$

$$\text{UCL} = \bar{p} + 3 * \sqrt{(\bar{p}(1-\bar{p}))/n}$$

$$\text{LCL} = \bar{p} - 3 * \sqrt{(\bar{p}(1-\bar{p}))/n}$$

The control limits will vary for each subgroup because the sample size n varies from group to group.

11.9 *g*-CHARTS

g-charts assume a binomial distribution and are usually used to chart the interval between rare events. A minimum of 25 observations is necessary to construct the chart.

$$g = \text{Interval between rare events}$$

$$\bar{g} = \frac{\Sigma g}{k} \text{ (Centerline)}$$

$$\text{UCL} = \bar{g} + 3 * \sqrt{\bar{g} * (\bar{g} + 1)}$$

$$\text{LCL} = \bar{g} - 3 * \sqrt{\bar{g} * (\bar{g} + 1)}$$

$$s \cong \sqrt{\bar{g} * (\bar{g} + 1)}$$

(Estimate of sigma for chart)

where

    k = The number of subgroups

Example: A company wishes to track the days between injuries to evaluate the presence of a pattern or trend. Twenty-five subgroups are evaluated (Table 11.8). Calculate the centerline, the upper and lower control limits, and construct a *g*-chart (Figure 11.11).

Table 11.8   Data for *g*-chart.

Subgroup	Days
1	0
2	17
3	15
4	12
5	29
6	17
7	17
8	10
9	11
10	14
11	25
12	15
13	47
14	16
15	9
16	22
17	16
18	7
19	11
20	29
21	11
22	17
23	37
24	13
25	16
Sum	**433**

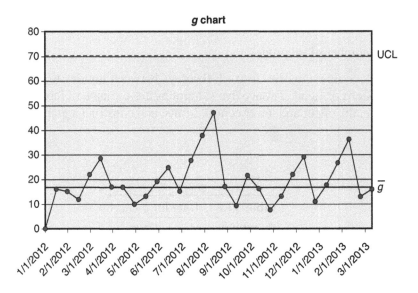

Figure 11.11 *g*-chart example.

$$\bar{g} = \frac{\Sigma g}{k} = \frac{433}{25} = 17.320$$

$$\text{UCL} = \bar{g} + 3 * \sqrt{\bar{g} * (\bar{g}+1)} = 17.320 + 3 * \sqrt{17.320 * (17.320+1)} = 70.759$$

$$\text{LCL} = \bar{g} - 3 * \sqrt{\bar{g} * (\bar{g}+1)} = 17.320 - 3 * \sqrt{17.320 * (17.320+1)} = -36.119,$$

which reverts to zero.

11.10 X AND mR (MOVING RANGE) CONTROL CHARTS

X and mR (moving range) control charts assume a normal distribution and are used with individual values. A minimum of 25 observations is necessary to construct the chart.

X chart	mR chart
X (Individual observation)	$\mathrm{mR} = \lvert X_2 - X_1 \rvert, \lvert X_3 - X_2 \rvert \ldots$
$\bar{X} = \dfrac{\Sigma X}{k}$ (Centerline)	$\overline{\mathrm{mR}} = \dfrac{\Sigma \mathrm{mR}}{k-1}$ (Centerline)
$\mathrm{UCL} = \bar{X} + d_2 * \overline{\mathrm{mR}}$	$\mathrm{UCL} = D_4 * \overline{\mathrm{mR}}$
$\mathrm{LCL} = \bar{X} - d_2 * \overline{\mathrm{mR}}$	$\mathrm{LCL} = 0$

$$ s = \frac{\overline{\mathrm{mR}}}{d_2} \text{ (Estimate of sigma)} $$

(See Appendix O, Control Chart Constants.)

where

   k = The number of subgroups

Example: A manager wants to evaluate a low-volume production process. Due to the cost of testing, one sample is measured per shift (Table 11.9). Calculate the centerline and the upper and lower control limits, and construct an X and mR chart (Figure 11.12).

$$ \bar{X} = \frac{\Sigma X}{k} = \frac{174.700}{25} = 6.988 $$

$$ \overline{\mathrm{mR}} = \frac{\Sigma \mathrm{mR}}{k-1} = \frac{7.200}{25-1} = 0.300 $$

$$ \mathrm{UCL} = \bar{X} + d_2 * \overline{\mathrm{mR}} = 6.988 + 1.128 * 0.300 = 7.326 $$

$$ \mathrm{LCL} = \bar{X} - d_2 * \overline{\mathrm{mR}} = 6.988 - 1.128 * 0.300 = 6.650 $$

$$ \mathrm{UCL} = D_4 * \overline{\mathrm{mR}} = 3.267 * 0.300 = 0.980 $$

$$ \mathrm{LCL} = 0 $$

Table 11.9 Data for X and mR chart.

Subgroup	X	mR
1	7.2	0.000
2	7.5	0.300
3	7.1	0.400
4	6.9	0.200
5	6.8	0.100
6	6.6	0.200
7	6.3	0.300
8	6.9	0.600
9	7.2	0.300
10	7.1	0.100
11	6.9	0.200
12	7.2	0.300
13	7.4	0.200
14	7.5	0.100
15	7.4	0.100
16	7.3	0.100
17	7.7	0.400
18	6.7	1.000
19	7.2	0.500
20	6.8	0.400
21	6.6	0.200
22	6.7	0.100
23	6.2	0.500
24	6.7	0.500
25	6.8	0.100
Sum	**174.700**	**7.200**

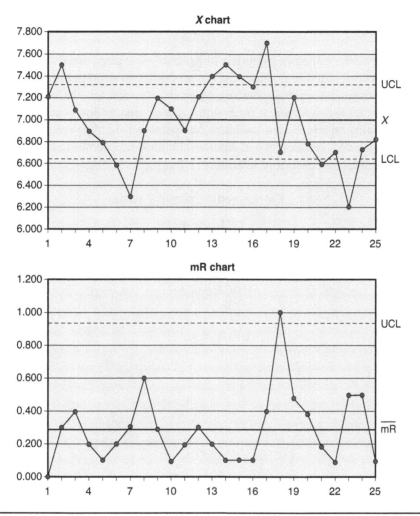

Figure 11.12 X and $\overline{mR}$ chart example.

11.11 PRE-CONTROL CHARTS

Pre-control charts, sometimes referred to as *stoplight charts*, are distribution free and are used to record individual values during production startup before there are enough data to use statistical control charting methods. Pre-control charts divide the part/process specification into three zones—red, green, and yellow—and plotting a data point. Depending on which zone the latest data point falls in, the process continues as is, is further monitored, or stopped and adjusted. Figure 11.13 demonstates the relationship between the zones and part print tolerance.

$$X = (\text{Centerline/nominal value})$$

$$\text{UTL} = X + \text{USL}$$

$$\text{LTL} = X - \text{LSL}$$

$$\text{UPCL} = X + \frac{\text{USL}}{2}$$

$$\text{LPCL} = X - \frac{\text{LSL}}{2}$$

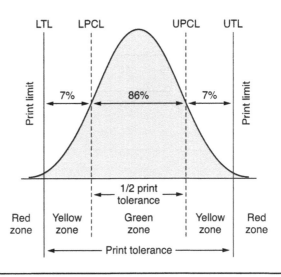

Figure 11.13   Assumptions for underlying pre-control.

where

X = Target value

UTL = Upper tolerance limit

LTL = Lower tolerance limit

USL = Allowed positive deviation from nominal

LSL = Allowed negative deviation from nominal

UPCL = Upper pre-control limit

LPCL = Lower pre-control limit

Interpretation rules:

- If five parts in a row fall between LPCL–UPCL (green zone), run the process
- If one part falls between the LTL-LPCL or UPCL-UTL (yellow zone), restart the count
- If two parts in a row fall between the LTL-LPCL or UPCL-UTL (yellow zone), adjust the process and restart the count
- If one part exceeds the LTL or UTL (red zone), adjust the process and restart the count
- If 25 parts in a row fall between the LPCL–UPCL (green zone), run the process, but at a reduced sampling frequency

Example: A machining process has a nominal dimension of 1.000" ± 0.010". Twenty-five subgroups are evaluated (Table 11.10). Construct a pre-control chart to monitor the initial production run (Figure 11.14).

$$\text{UTL} = X + \text{USL} = 1.000 + 0.010 = 1.010$$

$$\text{LTL} = X - \text{LSL} = 1.000 - 0.010 = 0.990$$

$$\text{UPCL} = X + \frac{\text{USL}}{2} = 1.000 + \frac{0.010}{2} = 1.005$$

$$\text{LPCL} = X - \frac{\text{LSL}}{2} = 1.000 - \frac{0.010}{2} = .995$$

Table 11.10   Data for pre-control chart.

Subgroup	X
1	1.003
2	1.001
3	1.004
4	1.004
5	0.993
6	1.002
7	0.999
8	0.999
9	1.013
10	0.997
11	1.002
12	1.003
13	1.002
14	1.002
15	1.000
16	1.003
17	0.994
18	0.988
19	0.994
20	0.998
21	1.001
22	0.996
23	1.002
24	1.000
25	0.997

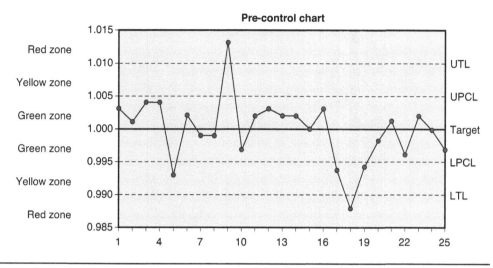

Figure 11.14   Pre-control chart example.

12

Process Capability

*P*rocess capability refers to the ability of a process to meet specifications set by the customer or designer. Process capability compares the output of an in-control process to the specification limits by using various capability indices. A capable process is one where almost all the measurements fall inside the specification limits (Figure 12.1).

12.1 PROCESS CAPABILITY FOR VARIABLES DATA

There are three potential process capability indices for variables data: C_r, C_p, and C_{pk}. Additionally, there are three process capability indices for variables data: Pr, P_p, and P_{pk}. The main difference between the "C" and "P" indices is that C_r, C_p, and C_{pk} are used with sample data to determine if the process is capable of meeting customer specifications (short term), whereas Pr, P_p, and P_{pk} are used with population data to determine if the process is capable of meeting customer specifications (long term).

These indices compare the process statistical control limits to the product specifications to determine the process capability. Figure 12.2 demonstrates the relationship between statistical control limits and product specifications.

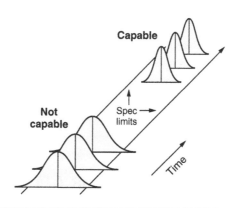

Figure 12.1　Capable process versus not capable process.
Source: Adapted from *Innovative Control Charting*, Figure 3.3. Used with permission.

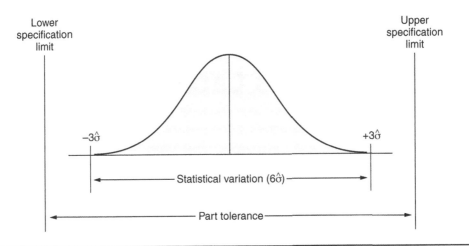

Figure 12.2 Relationship between statistical control limits and product specifications.

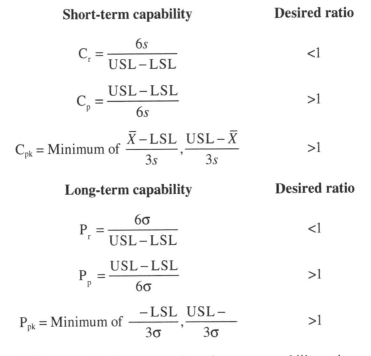

Figure 12.3 is a graphical representation of process capability ratios compared to the statistical control limits and part specification limits.

12.2 PROCESS CAPABILITY FOR ATTRIBUTES DATA

The process capability for attributes data is simply the process average. Ideally, the process average should be zero.

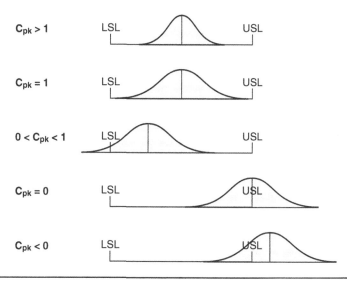

Figure 12.3 Graphical relationship of process capability ratios.

Table 12.1 Fall-out rate conversion table.

C_{pk}	Z	ppm	% defective	% yield
Given	3 * C_{pk}	Z table value * 2,000,000	ppm/1,000,000	1 – % defective
Z/3	Given	Z table value * 2,000,000	ppm/1,000,000	1 – % defective
Z/3	ppm/2,000,000 (Z table value)	Given	ppm/1,000,000	1 – % defective
Z/3	(%)/2 (Z table value)	(% defective) * 1,000,000	Given	1 – % defective
Z/3	(%)/2 (Z table value)	(% defective) * 1,000,000	1 – % yield	Given

12.3 INCREASING PROCESS CAPABILITY

There are three ways to increase process capability: decrease the process variation, center the process, or increase the product specification limits.

12.4 FALL-OUT RATES

There are often times when one needs to calculate the C_{pk}, Z, parts per million (ppm), or percentage defective. Table 12.1 gives the conversion formulas.

See Appendix B, Normal Distribution Probability Points—Area above Z.

Example: A process has a C_{pk} of 0.70. Calculate Z, ppm, percentage defective, and yield percentage.

$$Z = 3 * C_{pk} = 3 * 0.70 = 2.10$$

$$\text{ppm} = Z \text{ table value} * 2,000,000 = .0179 * 2,000,000 = 35,728$$

$$\% \text{ defective} = \text{ppm}/1,000,000 = 35,728/1,000,000 = .0358 \text{ or } 3.58\%$$

$$\% \text{ yield} = 1 - \% \text{ defective} = 1 - 3.58\% = 96.42\%$$

Example: A process has a Z value of 2.10. Calculate C_{pk}, ppm, percentage defective, and yield percentage.

$$C_{pk} = Z/3 = 2.1/3 = 0.70$$

$$\text{ppm} = Z \text{ table value} * 2,000,000 = .0179 * 2,000,000 = 35,728$$

$$\% \text{ defective} = \text{PPM}/1,000,000 = 35,728/1,000,000 = .0358 \text{ or } 3.58\%$$

$$\% \text{ yield} = 1 - \% \text{ defective} = 1 - 3.58\% = 96.42\%$$

Example: A process has a ppm of 35,728. Calculate Z, C_{pk}, ppm, percentage defective, and yield percentage.

$$C_{pk} = \text{ppm}/2,000,000 \ (Z \text{ table value}) = 35,728/2,000,000 = 0.0179 \rightarrow 2.10$$

$$C_{pk} = Z/3 = 2.1/3 = 0.70$$

$$\% \text{ defective} = \text{ppm}/1,000,000 = 35,728/1,000,000 = .0358 \text{ or } 3.58\%$$

$$\% \text{ yield} = 1 - \% \text{ defective} = 1 - 3.58\% = 96.42\%$$

Example: A process has a percentage defective of .0358%. Calculate Z, C_{pk}, ppm, and yield percentage.

$$Z = (\%)/2 \ Z \text{ table value} = .0358/2 = .0179 \rightarrow 2.10$$

$$C_{pk} = Z/3 = 2.1/3 = 0.70$$

$$\text{ppm} = \% \text{ defective} * 1,000,000 = .0358 * 1,000,000 = 35,800$$

$$\% \text{ yield} = 1 - \% \text{ defective} = 1 - .0358\% = 96.42\%$$

Example: A process has a yield percentage of 96.42%. Calculate Z, C_{pk}, ppm, and yield percentage.

$$Z = (\%)/2 \ Z \text{ table value} = .0358/2 = .0179 \rightarrow 2.10$$

$$C_{pk} = Z/3 = 2.1/3 = 0.70$$

$$\text{ppm} = \% \text{ defective} * 1,000,000 = .0358 * 1,000,000 = 35,800$$

$$\% \text{ defective} = 1 - \% \text{ yield} = 1 - 96.42\% = .0358\%$$

Note: You will notice some differences in the calculated values. These differences are due to rounding.

Appendix Q, Fall-Out Rates, is a chart that compares C_{pk} values, z-scores, one-tail probabilities, two-tail probabilities, one-tail ppm, and two-tail ppm.

12.5 DEFECTS PER MILLION OPPORTUNITIES (DPMO)

Defects per million opportunities (DPMO) refers to the number of defects interpolated to a million opportunities. To calculate the DPMO, we use the following formula:

$$DPMO = (1,000,000 * D)/(U * O)$$

where

D = Number of defects observed in the sample

U = Number of units in the sample

O = Opportunities per unit

Example: 1500 units have been inspected, with nine defects discovered. Each unit has the potential for three defects. Calculate the DPMO:

$$DPMO = (1,000,000 * D)/(U * O) = (1,000,000 * 9)(1500 * 3) = 2000$$

13

Acceptance Sampling

A*cceptance sampling* is used to assess the quality of a lot or batch of products based on a sample size, acceptance number, and desired quality level. Acceptance sampling has its advantages and limitations.

Advantages:

- Less damage due to handling (inspections)
- More economical than 100% inspection
- Less time than doing 100% inspection
- Especially useful if using destructive testing techniques

Limitations:

- Producer's and consumer's risk associated with sampling
- Sampling does not provide the complete picture of the true quality
- Quality can not be inspected into a product

13.1 C = 0 SAMPLING PLAN

$C = 0$ sampling plans are based on the premise of accepting the lot if zero defects are found during the inspection, and rejecting the lot if one or more defects are found during the inspection. Lots that are rejected may be subjected to further inspection or sorting, often at the expense of the supplier. $C = 0$ sampling plans provide more protection to the consumer, which is especially important when health and human welfare are involved.

13.2 AVERAGE OUTGOING QUALITY

The *average outgoing quality* (AOQ) is the maximum average outgoing quality over all possible levels of incoming quality for a given acceptance sampling plan. It can be calculated using the following formula for a $C = 0$ sampling plan:

$$AOQ = .3679\,[(1/n) - (1/N)]$$

where

n = Sample size

N = Lot size

Example: 19 samples were randomly selected and inspected from a lot of 150 pieces, with zero defects found. What is the AOQ?

$$\text{AOQ} = .3679\,[(1/n)-(1/N)] = .3679\,[(1/19)-(1/150)] = 0.169 \text{ or } 1.69\%$$

13.3 UPPER RISK LEVEL (CONFIDENCE STATEMENT)

To calculate the worst possible quality level for a lot that was inspected with zero defects being found, use the following formula:

$$p = -\ln(1-cl)\,/\,n$$

where

n = Sample size

p = Upper risk level

cl = Confidence level

Example: 19 samples were inspected, with zero defects being found. What is the upper risk level at the 95% confidence level?

$$p = -\ln(1-cl)\,/\,n = -\ln(1-.95)/19 = 0.1577 \text{ or } 15.77\%$$

We are 95% confident the true defect rate is between 0 and 15.77%. Additionally, it is 95% probable of finding one or more defects in the sample.

13.4 SAMPLE SIZE REQUIRED TO FIND PERCENTAGE DEFECTIVE WITH A GIVEN CONFIDENCE LEVEL

To calculate the sample size required to detect an upper risk level with a given level of confidence, the formula is

$$n = -\ln(1-cl)\,/\,p$$

where

n = Sample size

p = Upper risk level

cl = Confidence level

Example: A lot is suspected of having a defect rate of 1%. What is the sample size required to be 95% confident that we can detect at least one defect?

$$n = -\ln(1-cl)/p = -\ln(1-.95)/0.01 = 299.6$$

rounded up to the next integer is 300. Three hundred samples must be inspected to be 95% confident that at least one or more defects will be found when the lot contains 1% defects.

14

ANOVA

nalysis of variance (ANOVA) is a collection of statistical models used to analyze the differences between three or more groups of means and their associated variation among and between groups. In the ANOVA setting, the observed
variance in a particular variable is partitioned into components attributable to different sources of variation. In its simplest form, ANOVA provides a statistical test of
whether or not the means of three or more groups are equal, and therefore generalizes
the *t*-test to more than two groups. Doing multiple two-sample *t*-tests would result in
an increased chance of committing a type I error. For this reason, ANOVAs are useful
in comparing three or more means for statistical significance (http://en.wikipedia.org/
wiki/Analysis_of_variance).

14.1 ONE-WAY ANOVA

In statistics, *one-way analysis of variance* is a technique used to compare means of two
or more samples (using the *F*-distribution). This technique can be used only for numeri-
cal data.

The ANOVA tests the null hypothesis that samples in two or more groups are drawn
from populations with the same mean values. To do this, two estimates are made of the
population variance. These estimates rely on various assumptions. The ANOVA produces an *F*-statistic, the ratio of the variance calculated between the means to the variance within the samples. If the group means are drawn from populations with the same
mean values, the variance between the group means should be lower than the variance
of the samples, following the central limit theorem. A higher ratio therefore implies that
the samples were drawn from populations with different mean values.

Typically, however, the one-way ANOVA is used to test for differences between at
least three groups, since the two-group case can be covered by a *t*-test. When there are
only two means to compare, the *t*-test and the *F*-test are equivalent. A one-way ANOVA
summary table is shown in Table 14.1.

The results of a one-way ANOVA can be considered reliable as long as the following assumptions are met:

- Response variable residuals are normally distributed.

- Samples are independent.

Table 14.1   One-way ANOVA summary table.

Source of variation	SS	df	MS	F calculated	F critical
Between groups					
Within groups (error)					
Total					

- Variances of populations are equal.

- Data are interval or nominal.

(http://en.wikipedia.org/wiki/One-way_ANOVA)

The hypothesis statements for a one-way ANOVA are as follows:

$$H_0 : \mu_1 = \mu_2 = \mu_3 \ ... \ \mu_k \quad \text{(Null hypothesis)}$$

$$H_A : \text{The means are not equal (Alternative hypothesis)}$$

$$SS_{\text{Total}} = \Sigma X_N^2 - \frac{(\Sigma X_N)^2}{N}$$

$$SS_{\text{Between}} = \frac{(\Sigma X_1)^2}{n_1} + \frac{(\Sigma X_2)^2}{n_2} + ... + \frac{(\Sigma X_k)^2}{n_k} - \frac{(\Sigma X_N)^2}{N}$$

$$SS_{\text{Within}} = SS_{\text{Total}} - SS_{\text{Between}}$$

$$df_{\text{Total}} = N - 1$$

$$df_{\text{Between}} = k - 1$$

$$df_{\text{Within}} = N - k$$

$$MS_{\text{Between}} = \frac{SS_{\text{Between}}}{df_{\text{Between}}}$$

$$MS_{\text{Within}} = \frac{SS_{\text{Within}}}{df_{\text{Within}}}$$

$$F = \frac{MS_{\text{Between}}}{MS_{\text{Within}}}$$

where

N = Total number of observations

K = Number of treatments

$F_{\text{Critical}} = \alpha$, df, df values of the F-distribution (See Appendix G, Percentages of the F-Distribution)

Example: An engineer is unsure if a process is sensitive to a pressure setting. The engineer decides to conduct an ANOVA using three pressure settings: 100 psi, 110 psi, and

120 psi. The outputs are recorded below. We wish to test whether a change in pressure has an effect on the output with an α of 0.05.

Pressure (psi)		
100	**110**	**120**
8	5	2
9	4	4
8	5	5
7	7	2
9	5	7

$$SS_{\text{Total}} = \Sigma X_N^2 - \frac{(\Sigma X_N)^2}{N} = 577 - \frac{87^2}{15} = 72.4$$

$$SS_{\text{Between}} = \frac{(\Sigma X_1)^2}{n_1} + \frac{(\Sigma X_2)^2}{n_2} + \frac{(\Sigma X_3)^2}{n_3} = \frac{41^2}{5} + \frac{26^2}{5} + \frac{20^2}{5} = 46.8$$

$$SS_{\text{Within}} = SS_{\text{Total}} - SS_{\text{Between}} = 72.4 - 46.8 = 25.6$$

$$\text{df}_{\text{Total}} = N - 1 = 15 - 1 = 14$$

$$\text{df}_{\text{Between}} = k - 1 = 3 - 1 = 2$$

$$\text{df}_{\text{Within}} = N - k = 15 - 3 = 12$$

$$MS_{\text{Between}} = \frac{SS_{\text{Between}}}{\text{df}_{\text{Between}}} = \frac{46.8}{2} = 23.4$$

$$MS_{\text{Within}} = \frac{SS_{\text{Within}}}{\text{df}_{\text{Within}}} = \frac{25.6}{12} = 2.1$$

$$F = \frac{MS_{\text{Between}}}{MS_{\text{Within}}} = \frac{23.4}{2.1} = 10.97$$

The results are shown in Table 14.2. Since F calculated is greater than F critical at the 0.05 α (alpha) level, there is sufficient evidence to reject the null hypothesis $H_0 : \mu_1 = \mu_2 = \mu_3$ in favor of the alternative hypothesis that H_A : The means are not equal.

Table 14.2 One-way ANOVA summary data table.

Source of variation	SS	df	MS	F calculated	F critical
Between groups	46.8	2	23.4	10.97	3.89
Within groups (error)	25.6	12	2.1		
Total	72.4	14			

14.2 TWO-WAY ANOVA

In statistics, the *two-way analysis of variance* (ANOVA) test is an extension of the one-way ANOVA test. It examines the influence of different categorical independent variables on one dependent variable. While the one-way ANOVA measures the significant effect of one independent variable, the two-way ANOVA is used when there are more than one independent variable and multiple observations for each independent variable. The two-way ANOVA can not only determine the main effect of the contributions of each independent variable, but also identifies whether there is a significant interaction effect between the independent variables. A two-way ANOVA summary table is shown in Table 14.3.

As with other parametric tests, we make the following assumptions when using two-way ANOVA:

- Response variable residuals are normally distributed.

- Samples are independent.

- Variances of populations are equal.

- Data are interval or nominal.

(http://en.wikipedia.org/wiki/Two-way_analysis_of_variance)

The hypothesis statements for a two-way ANOVA are as follows:

$$H_0 \text{ (Rows)} : \mu_1 = \mu_2 = \mu_3 \ldots \mu_k \text{ (Null hypothesis)}$$

$$H_0 \text{ (Columns)} : \mu_1 = \mu_2 = \mu_3 \ldots \mu_k \text{ (Null hypothesis)}$$

$$H_0 \text{ (Interactions)} : \mu_1 = \mu_2 = \mu_3 \ldots \mu_k \text{ (Null hypothesis)}$$

$$H_A : \text{The means are not equal (Alternative hypothesis)}$$

$$SS_{\text{Total}} = \Sigma X_N^2 - \frac{(\Sigma X_N)^2}{N}$$

$$SS_{\text{Between}} = \frac{(\Sigma X_1)^2}{n_1} + \frac{(\Sigma X_2)^2}{n_2} + \ldots + \frac{(\Sigma X_k)^2}{n_k} - \frac{(\Sigma X_N)^2}{N}$$

Table 14.3 Two-way ANOVA summary table.

Source of variation	SS	df	MS	F calculated	F critical
Rows					
Columns					
Rows × Columns (interaction)					
Within (error)					
Total					

Please note that this formula requires data from each individual cell. For example, C1-R1, C2-R1, C3-R1, C1-R2, C2-R2, C3-R2.

$$SS_{\text{Within}} = SS_{\text{Total}} - SS_{\text{Between}}$$

$$SS_{\text{Row}} = \Sigma \frac{\left(\Sigma_{\text{For each row}}\right)^2}{n_{\text{For each row}}} - \frac{\left(\Sigma X_N\right)^2}{N}$$

$$SS_{\text{Column}} = \Sigma \frac{\left(\Sigma_{\text{For each column}}\right)^2}{n_{\text{For each column}}} - \frac{\left(\Sigma X_N\right)^2}{N}$$

$$SS_{\text{Interaction}} = SS_{\text{Between}} - SS_{\text{Row}} - SS_{\text{Column}}$$

$$df_{\text{Total}} = N - 1$$

$$df_{\text{Row}} = r - 1$$

$$df_{\text{Column}} = c - 1$$

$$df_{\text{Row} \times \text{Column}} = (r - 1)(c - 1)$$

$$df_{\text{Within}} = N - rc$$

$$MS_{\text{Within}} = \frac{SS_{\text{Within}}}{df_{\text{Within}}}$$

$$MS_{\text{Row}} = \frac{SS_{\text{Row}}}{df_{\text{Row}}}$$

$$MS_{\text{Column}} = \frac{SS_{\text{Column}}}{df_{\text{Column}}}$$

$$MS_{\text{Row} \times \text{Column}} = \frac{SS_{\text{Row} \times \text{Column}}}{df_{\text{Row} \times \text{Column}}}$$

$$F_{\text{Row}} = \frac{MS_{\text{Row}}}{MS_{\text{Within}}}$$

$$F_{\text{Column}} = \frac{MS_{\text{Column}}}{MS_{\text{Within}}}$$

$$F_{\text{Row} \times \text{Column}} = \frac{MS_{\text{Row} \times \text{Column}}}{MS_{\text{Within}}}$$

where

N = Total number of observations

r = Number of rows

c = Number of columns

$F_{\text{Critical}} = \alpha$, df, df values of the F-distribution (See Appendix G, Percentages of the F-Distribution)

Example: An engineer is unsure whether a process is sensitive to a change in presses and dwell time. The engineer decides to conduct an ANOVA using three dwell times: 1.0 second, 1.5 seconds, and 2.5 seconds using press 1 and press 2. The outputs are recorded below. We wish to test whether a change in pressure has an effect on the output with an α of 0.05.

		Dwell time	
	1.0	**1.5**	**2.5**
Press 1	8	5	2
	9	4	4
	8	5	5
Press 2	7	7	2
	9	5	7
	7	6	5

$$SS_{\text{Total}} = \Sigma X_N^2 - \frac{(\Sigma X_N)^2}{N} = 687 - \frac{105^2}{18} = 74.5$$

$$SS_{\text{Between}} = \frac{(\Sigma X_1)^2}{n_1} + \frac{(\Sigma X_2)^2}{n_2} + \ldots + \frac{(\Sigma X_k)^2}{n_k} - \frac{(\Sigma X_N)^2}{N}$$

$$SS_{\text{Between}} = \frac{25^2}{3} + \frac{14^2}{3} + \frac{11^2}{3} + \frac{23^2}{3} + \frac{18^2}{3} + \frac{14^2}{3} - \frac{105^2}{18} = 51.2$$

$$SS_{\text{Within}} = SS_{\text{Total}} - SS_{\text{Between}} = 74.5 - 51.2 = 23.3$$

$$SS_{\text{Row}} = \Sigma \frac{(\Sigma_{\text{For each row}})^2}{n_{\text{For each row}}} - \frac{(\Sigma X_N)^2}{N} = \frac{50^2}{9} + \frac{55^2}{9} - \frac{105^2}{18} = 1.4$$

$$SS_{\text{Column}} = \Sigma \frac{(\Sigma_{\text{For each column}})^2}{n_{\text{For each column}}} - \frac{(\Sigma X_N)^2}{N} = \frac{48^2}{6} + \frac{32^2}{6} + \frac{25^2}{6} - \frac{105^2}{18} = 46.3$$

$$SS_{\text{Interaction}} = SS_{\text{Between}} - SS_{\text{Row}} - SS_{\text{Column}} = 51.2 - 1.4 - 46.3 = 3.5$$

$$df_{\text{Total}} = N - 1 = 18 - 1 = 17$$

$$df_{\text{Row}} = r - 1 = 2 - 1 = 1$$

$$df_{\text{Column}} = c - 1 = 3 - 1 = 2$$

$$df_{\text{Row} \times \text{Column}} = (r - 1)\,(c - 1) = (2 - 1)\,(3 - 1) = 2$$

$$df_{\text{Within}} = N - rc = 18 - 2 * 3 = 12$$

Table 14.4   Two-way ANOVA summary data table.

Source of variation	SS	df	MS	F calculated	F critical
Rows	1.4	1	1.4	0.74	4.75
Columns	46.3	2	23.2	12.21	3.89
Rows × Columns (interaction)	3.5	2	1.8	0.95	3.89
Within (error)	23.3	12	1.9		
Total	**74.5**	**17**			

$$MS_{\text{Within}} = \frac{SS_{\text{Within}}}{df_{\text{Within}}} = \frac{23.3}{12} = 1.9$$

$$MS_{\text{Row}} = \frac{SS_{\text{Row}}}{df_{\text{Row}}} = \frac{1.4}{1} = 1.4$$

$$MS_{\text{Column}} = \frac{SS_{\text{Column}}}{df_{\text{Column}}} = \frac{46.3}{2} = 23.2$$

$$MS_{\text{Row} \times \text{Column}} = \frac{SS_{\text{Row} \times \text{Column}}}{df_{\text{Row} \times \text{Column}}} = \frac{3.5}{2} = 1.8$$

$$F = \frac{MS_{\text{Row}}}{MS_{\text{Within}}} = \frac{1.4}{1.9} = 0.74$$

$$F = \frac{MS_{\text{Column}}}{MS_{\text{Within}}} = \frac{23.2}{1.9} = 12.21$$

$$F = \frac{MS_{\text{Row} \times \text{Column}}}{MS_{\text{Within}}} = \frac{1.8}{1.9} = 0.95$$

Results are shown in Table 14.4.

Results for rows: Since F calculated is less than F critical at the 0.05 α (alpha) level, there is insufficient evidence to reject the null hypothesis that H_0 (Rows) : $\mu_1 = \mu_2 = \mu_3 \ldots \mu_k$.

Results for columns: Since F calculated is greater than F critical at the 0.05 α (alpha) level, there is sufficient evidence to reject the null hypothesis H_0 (Columns) : $\mu_1 = \mu_2 = \mu_3 \ldots \mu_{k2}$ in favor of the alternative hypothesis that H_A : The means are not equal.

Results for interactions: Since F calculated is less than F critical at the 0.05 α (alpha) level, there is insufficient evidence to reject the null hypothesis that H_0 (Interactions) : $\mu_1 = \mu_2 = \mu_3 \ldots \mu_k$.

15

The Reliability Bathtub Curve

eliability is the probability that an item will perform its intended function for a specified interval (time or cycles) under stated environmental conditions. The reliability bathtub curve helps us to understand the concept of reliability (see Figure 15.1).

Over the life of a complex system, three distinct failure rate phases become apparent. The first phase is the *infant mortality* phase, which has a decreasing failure rate (see Figure 15.2). The second phase is the *random failure* phase, which has a constant failure rate (see Figure 15.3). The third phase is the *wear-out* phase, which has an increasing failure rate (see Figure 15.4). The wear-out phase is more predominant in mechanical systems than in electronic systems.

Infant mortality failures are generally the result of manufacturing errors that are not caught in inspection prior to burn-in or placing into service. Failures resulting from time/stress-dependent errors may occur in this period.

Random failures and wear-out failures are generally a factor of design, that is, improper materials, excessive load, and so on.

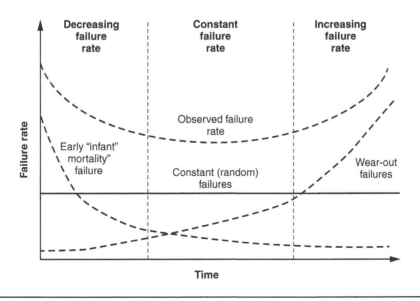

Figure 15.1 Reliability bathtub curve.

No distinct break-off from infant mortality to random to wear-out failures has been established. Random failures can occur anywhere in the three periods, as can infant mortality failures. A failure caused by a cold solder joint may occur well into the service life, but this is really an infant mortality type failure. Wear-out of mechanical parts also begins the moment the product is put into service.

The probability distributions occurring most often during infant mortality are the Weibull, gamma, and decreasing exponential. Probability distributions of value in the constant failure rate phase are the exponential and Weibull. The wear-out phase generally follows the normal or Weibull distributions.

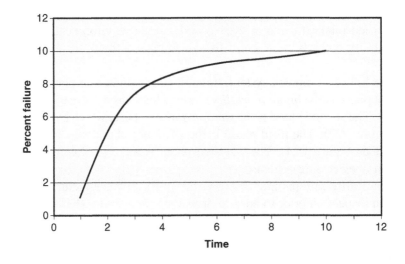

Figure 15.2   Decreasing failure rate.

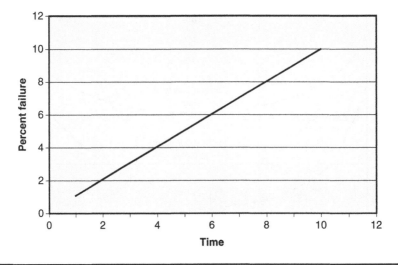

Figure 15.3   Constant failure rate.

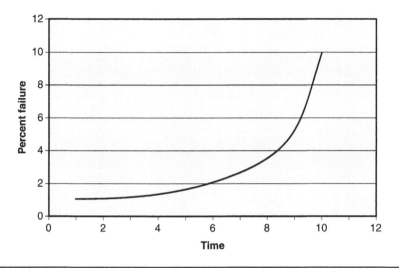

Figure 15.4   Increasing failure rate.

16
Availability

Availability is the probability that a system is operating satisfactorily at any moment in time when used under stated conditions. There are three types of availability:

- Inherent availability
- Achieved availability
- Operational availability

16.1 INHERENT AVAILABILITY

Inherent availability is the ideal state and is a function of the *mean time before failure* (MTBF) (reliability) and the *mean time to repair* (MTTR) (maintainability).

$$A_i = \frac{MTBF}{MTBF + MTTR}$$

where

A_i = Inherent availability

MTBF = Mean time before failure

MTTR = Mean time to repair

Example: A repairable system has an MTBF of 2000 hours. The same system has an MTTR of eight hours. What is the inherent availability of the system?

$$A_i = \frac{MTBF}{MTBF + MTTR} = \frac{2000}{2000 + 8} = .996 \text{ or } 99.6\%$$

16.2 ACHIEVED AVAILABILITY

Achieved availability includes the measure of preventive and corrective maintenance. Achieved availability is calculated using the *mean time between maintenance actions* (MTBMA) and *mean maintenance action time* (MMAT).

$$MMAT = \frac{Fc\overline{M}ct + Fp\overline{M}pt}{Fc + Fp}$$

where

MMAT = Mean maintenance action time

Fc = Number of corrective actions per 1000 hours

Fp = Number of preventive actions per 1000 hours

$Fc\overline{M}ct$ = Mean active corrective maintenance time

$Fp\overline{M}pt$ = Mean active preventive maintenance time

Example: A repairable system has one corrective action per 1000 hours, two preventive actions per 1000 hours, a mean active corrective maintenance time of 8.5 hours, and mean active preventive maintenance time of two hours. What is the mean maintenance action time?

$$MMAT = \frac{Fc\overline{M}ct + Fp\overline{M}pt}{Fc + Fp} = \frac{(1*8.5) + (2*2)}{1 + 2} = 4.17 \text{ hours}$$

$$A_a = \frac{MTBMA}{MTBMA + MMAT}$$

where

A_a = Achieved availability

MTBMA = Mean time between maintenance actions

MMAT = Mean maintenance action time

The system described above has an MTBMA of 1000 hours. What is the achieved availability of the system?

$$A_a = \frac{MTBMA}{MTBMA + MMAT} = \frac{1000}{1000 + 4.17} = .996 \text{ or } 99.6\%$$

16.3 OPERATIONAL AVAILABILITY

Operational availability includes the measures of inherent availability as well as achieved availability, and includes administrative and logistical downtime.

$$A_o = \frac{\text{MTBMA}}{\text{MTBMA} + \text{MDT}}$$

where

A_o = Operational availability

MTBMA = Mean time between maintenance actions

MDT = Mean downtime

Example: A repairable system has an MTBMA of 1000 hours. The same system has an MTD of eight hours. What is the operational availability of the system?

$$A_o = \frac{\text{MTBMA}}{\text{MTBMA} + \text{MDT}} = \frac{1000}{1000 + 8} = .992 \text{ or } 99.2\%$$

17

Exponential Distribution

T he *exponential distribution* is one of the most commonly used distributions in reliability and is generally used to predict the probability of survival to time *t* (Figure 17.1).

The hazard function for the exponential distribution is λ and is constant throughout the function. Therefore, the exponential distribution should only be used for reliability prediction during the constant failure rate or random failure phase of operation. Some unique characteristics of the exponential distribution include:

- The mean and standard deviation are equal.

- 63.21% of all of the values fall below the mean value, which translates into only a 36.79% chance of surviving past the MTBF.

- Reliability, as time *t* approaches zero, approaches one as a limit.

The reliability for a given time *t* during the constant failure period can be calculated with the following formula:

$$R_{(t)} = e^{-\lambda t}$$

where

e = The base of natural logarithms 2.718281828 . . . .

λ = Failure rate

t = Time (Note: Cycles [c] may be substituted for time [t])

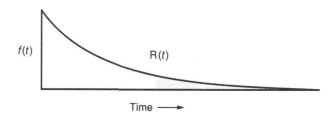

Figure 17.1   Exponential distribution.

Example: The equipment in a packaging plant has a failure rate of 0.001/hr (MTBF = 1000 hours). What is the probability of operating for a period of 500 hours without failure?

$$R_{(500)} = e^{-0.01(500)} = 0.6065 \text{ or } 60.65\%$$

The λ or MTBF do not need to be a function of time in hours. The characteristics of time or usage can be such units as cycles rather than hours. In this case *mean cycles before failure* (MCBF) would be the appropriate measure.

Example: One cycle of a molding machine produces 24 parts. A study of this machine predicted an MCBF of 15,000 cycles ($\lambda = 0.0000667$/cycle). What is the probability of operating 16,000 cycles without failure?

$$R_{(16,000)} = e^{-0.0000667(16,000)} = .3442 \text{ or } 34.42\%$$

$\lambda = \dfrac{1}{\theta}$ (Recall that the failure rate is the reciprical of the mean time before failure)

where

   λ = Failure rate

   θ = Estimate of MTBF/MCBF

Example: An item fails on the average of once every 10,000 hours. What is the probability of survival for 2000 hours?

$$\lambda = \frac{1}{\theta} = \frac{1}{10,000} = 0.0001$$

$$R_{(2000)} = e^{-0.0001(2000)} = 0.8187 \text{ or } 81.87\%$$

An interesting aspect of prediction during the random failure period is that the probability of functioning for a given time period (t) is totally independent of previous operation. Therefore, as long as the operation remains in random failure mode, the probability of failure remains the same.

18

Censoring and MTBF and MCBF Calculations

At times, several items are tested simultaneously, and the results are combined to estimate the mean time before failure/mean cycles before failure (MTBF/MCBF). When performing tests of this type, two common test results occur: type I and type II censoring. Note that the MTBF/MCBF (θ) is the reciprocal of the failure rate (λ).

18.1 TYPE I CENSORING

This is when a total of n items are placed on test. Each test stand operates each test item a given number of hours or cycles. As test items fail, they are replaced. The test time is defined in advance, so the test is said to be *truncated* after the specified number of hours or cycles. A time- or cycle-truncated test is called *type I censoring*. The formula for type I censoring is:

$$\theta = \frac{nt}{r}$$

where

   θ = Estimate of MTBF/MCBF

   n = The number of items on test

   t = Total test time or cycles

   r = The number of failures occurring during the test

$$\lambda = \frac{1}{\theta}$$

where

   λ = Failure rate

   θ = Estimate of MTBF/MCBF

Example: Twelve items were placed on test for 50,000 cycles each (600,000 total cycles). As units failed, they were repaired or replaced with new units. During the test, nine units failed. What is the estimate of the MCBF?

$$\theta = \frac{nt}{r} = \frac{12*50,000}{9} = 66,666.67 \text{ cycles}$$

What is the failure rate?

$$\lambda = \frac{1}{\theta} = \frac{1}{66,666.67} = 0.000015$$

18.1 TYPE II CENSORING

Type II censoring is where *n* items are placed on test one at a time, and the test is truncated after a total of *r* failures occurs. As items fail, they are not replaced or repaired.

$$\theta = \frac{\Sigma t_f + (n-r)t_t}{r}$$

where

   θ = Estimate of MTBF/MCBF

   n = The number of items on test

   r = The number of failures occurring during the test

   t_f = Total test failure time or cycles

   t_t = Time or cycles of last failure

Example: A total of 20 items are placed on test. The test is to be truncated when the fourth failure occurs. The failures occur at the following number of hours into the test.

Failure #	Test hour
1	317
2	736
3	1032
4	1335
Total	**3420**

The remaining 16 items were in good operating order when the test was truncated at the fourth failure. What is the estimate of the MTBF?

$$\theta = \frac{\Sigma t_f + (n-r)t_t}{r} = \frac{3420 + 16(1335)}{4} = 6195 \text{ hours}$$

What is the failure rate?

$$\lambda = \frac{1}{\theta} = \frac{1}{6195} = .0.00016$$

19

Confidence Intervals for MTBF/MCBF

The calculation of the MTBF/MCBF (θ) is like $\bar{X}$ and s, a point estimate. An improvement in this estimate in the form of a confidence interval is required in some cases. Confidence intervals allow a band or interval to be put around the point estimate, which adds more meaning to the estimate. For example, a 90% confidence interval means that 90% of the intervals calculated in this manner from sample data will contain the true (but unknown) mean, while 10% will not. At 95% confidence, 95% of the intervals calculated in this manner from sample data will contain the true (but unknown) mean, while 5% will not.

The higher the level of confidence, the wider the confidence interval will be. When assuming an exponential distribution (constant failure rate) the chi-squared distribution can be used to calculate these confidence intervals.

In reliability testing, to estimate MTBF/MCBF (θ), two situations exist. The first is testing for a given period of time/cycles (type I censoring); the second case is testing to a predetermined number of failures (type II censoring). While the differences are in test design, there is also a difference in the calculation of the lower confidence limit based on the test method used.

19.1 TESTING TO A PREDETERMINED TIME/CYCLES

When testing to a predetermined time/cycles the following equations are used.

Lower one-sided confidence limit:

$$\frac{2T}{\chi^2_{\alpha,2r}} \leq \theta$$

Two-sided confidence limit:

$$\frac{2T}{\chi^2_{\alpha/2,2r}} \leq \theta \leq \frac{2T}{\chi^2_{1-\alpha/2,2r}}$$

where

T = Test time

α = Level of risk (1 – Confidence)

r = The number of failures

$\chi^2_{\alpha,2r}$, $\chi^2_{\alpha/2,2r}$, and $\chi^2_{1-\alpha/2,2r}$ = Distribution of the chi-square value for a given confidence level for r degrees of freedom (see Appendix F, Distribution of the Chi-Square)

Example: Several units are tested until the fourth failure. Total test time when the fourth failure occurs is 5000 hours (θ = 1250 hours). Calculate the one-sided lower confidence interval at 90% confidence (α = 0.10), and calculate the two-sided confidence interval at 90% confidence (α = 0.10, $\alpha/2$ = 0.05, 1 – $\alpha/2$ = 0.95).

One-sided interval:

$$\frac{2T}{\chi^2_{\alpha,2r}} \le \theta$$

$$\frac{2*5000}{13.362} \le \theta$$

748.39 $\le \theta$ with 90% confidence

Two-sided interval:

$$\frac{2T}{\chi^2_{\alpha/2,2r}} \le \theta \le \frac{2T}{\chi^2_{1-\alpha/2,2r}}$$

$$\frac{2*5000}{15.507} \le \theta \le \frac{2*5000}{2.733}$$

644.87 $\le \theta \le$ 3658.98 with 90% confidence

19.2 TESTING TO A PREDETERMINED NUMBER OF FAILURES

When testing for a predetermined time, the following equations are used:

Lower one-sided confidence interval:

$$\frac{2T}{\chi^2_{\alpha,2r+2}} \le \theta$$

Two-sided confidence interval:

$$\frac{2T}{\chi^2_{\alpha/2,2r+2}} \leq \theta \leq \frac{2T}{\chi^2_{1-\alpha/2,2r}}$$

where

T = Test time

α = Level of risk (1 – Confidence)

r = The number of failures

$\chi^2_{\alpha,2r+2}$, $\chi^2_{\alpha/2,2r+2}$, and $\chi^2_{1-\alpha/2,2r}$ = Distribution of the chi-square value for a given confidence level for r degrees of freedom (see Appendix F, Distribution of the Chi-Square)

Example: Several units are tested for a total of 3000 hours. When testing was stopped after 3000 hours, it was noted that three failures had occurred during the testing (θ = 1000 hours). Calculate the one-sided lower confidence interval at 90% confidence (α = 0.10), and calculate the two-sided confidence interval at 90% confidence (α = 0.10, $\alpha/2$ = 0.05, 1 – $\alpha/2$ = 0.95).

One-sided interval:

$$\frac{2T}{\chi^2_{\alpha,2r+2}} \leq \theta$$

$$\frac{2*3000}{13.362} \leq \theta$$

$449.03 \leq \theta$ with 90% confidence

Two-sided confidence interval:

$$\frac{2T}{\chi^2_{\alpha/2,2r+2}} \leq \theta \leq \frac{2T}{\chi^2_{1-\alpha/2,2r}}$$

$$\frac{2*3000}{15.507} \leq \theta \leq \frac{2*3000}{1.635}$$

$386.9 \leq \theta \leq 3669.7$ with 90% confidence

19.3 FAILURE-FREE TESTING

When no failures occur in a test for a predetermined amount of time, the one-sided lower limit calculation is simplified to:

One-sided confidence limit:

$$\frac{nT}{\ln(\alpha)} \leq \theta$$

where

T = Test time

α = Level of risk (1 – Confidence)

n = Sample size

Example: Fifty units are tested without failure for 100 hours. What is the 95% lower confidence limit for θ?

$$\frac{nT}{\ln(\alpha)} \leq \theta$$

$$\frac{50 * 100}{\ln(0.05)} \leq \theta$$

$$1669.04 \leq \theta$$

20

Nonparametric and Related Test Designs

20.1 CALCULATING RELIABILITY IN ZERO-FAILURE SITUATIONS

When calculating the lower confidence limit for reliability when no failures are to be allowed during the testing, the situation is simplified by the use of the following equation:

$$R_L = \alpha^{1/n}$$

where

n = Sample size

α = Level of risk (1 – Confidence)

Example: Fifty units are tested without failure. What is the 95% confidence level of reliability?

$$R_L = \alpha^{1/n} = 0.05^{1/50} = .9418 \text{ or } 94.18\%$$

21

Sample Size, Reliability, and Confidence Determination

21.1 SAMPLE SIZE DETERMINATION BASED ON CONFIDENCE AND RELIABILITY WITH ZERO FAILURES ALLOWED

There are often times when the question arises of how many samples are necessary to perform a test without a failure to demonstrate both confidence and reliability. The following formula can perform this calculation:

$$n = \frac{\ln(1-C)}{\ln(R)}$$

where

n = Sample size

C = Confidence level

R = Reliability

Example: What is the sample size required to perform a test without failure to be 95% confident the part is 99% reliable?

$$n = \frac{\ln(1-C)}{\ln(R)} = \frac{\ln(1-.95)}{\ln(.99)} = 298.07 \text{ (rounded up to the next integer, 299)}$$

To be 95% confident the part is 99% reliable, 299 parts must be tested without failure.

21.2 RELIABILITY ESTIMATE WHEN SAMPLE SIZE IS PROVIDED

When the sample is provided, and tests are performed without a failure, with a given confidence, we can calculate the reliability with the following formula:

$$R = (1-C)^{1/(n+1)}$$

where

n = Sample size

C = Confidence level

R = Reliability

Example: What is the reliability when tests were performed without failure using 32 samples, and we require 90% confidence?

$$R = (1-C)^{1/(n+1)} = (1-0.90)^{1/(32+1)} = .9326 \text{ or } 93.26\%$$

21.3 SAMPLE SIZE CALCULATION WITH FAILURES ALLOWED

There are often times when the question arises of how many samples are necessary to perform a test with r number of failures to demonstrate both confidence and reliability. The following formula can perform this calculation:

$$n = \frac{0.5 \times \chi^2_{(1-C,2(r+1))}}{1-R}$$

where

n = Sample size

r = Number of failures

C = Confidence level

R = Reliability

$\chi^2_{(1-C,2(r+1))}$ = Distribution of the chi-square value for a given confidence level for r degrees of freedom (see Appendix F, Distribution of the Chi-Square)

Example: What is the sample size required to ensure we are 90% confident that we are 90% reliable when three failures occur?

$$n = \frac{0.5 \times \chi^2_{(1-C,2(r+1))}}{1-R} = \frac{0.5 \times \chi^2_{(1-0.90,2(3+1))}}{1-R} = \frac{0.5 \times 13.362}{1-0.90} = 66.81$$

rounded up to the next whole integer, 67 samples

21.4 RELIABILITY ESTIMATE WHEN SAMPLE SIZES ARE SPECIFIED

When the sample is provided and tests are performed with r number of failures, with a given confidence, we can calculate the reliability with the following formula:

$$R = 1 - \frac{0.5 \times \chi^2_{(1-C,2(r+1))}}{n}$$

where

n = Sample size

r = Number of failures

C = Confidence level

R = Reliability

$\chi^2_{(1-C,2(r+1))}$ = Distribution of the chi-square value for a given confidence level for r degrees of freedom (see Appendix F, Distribution of the Chi-Square)

Example: What is the reliability necessary to ensure we are 90% confident when 69 samples were tested and three failures occur?

$$R = 1 - \frac{0.5 \times \chi^2_{(1-C,2(r+1))}}{n} = 1 - \frac{0.5 \times \chi^2_{(1-0.90,2(3+1))}}{69} = 1 - \frac{0.5 \times 13.362}{69} = .9032$$

or 90.32% reliable

22

Wear-Out Distribution

In mechanical reliability applications, wear-out modes are important contributors to the failure patterns of complex equipment. The wear-out life for common components can be determined by testing and/or historical data. The standard deviation of wear-out failures can be reasonably approximated with a sample of 30 to 50 pieces, and may be considered known, rather than estimated from sample data, to simplify the calculation. If the standard deviation is not known, the one-sided statistical tolerance limit *K* for a normal distribution can be used.

22.1 WEAR-OUT DISTRIBUTION—STANDARD DEVIATION KNOWN

The intent of this application is to determine the failure distribution and to replace components before they enter the wear-out phase with its rapidly increasing failure rate. To calculate the lower limit of the wear-out phase with the known standard deviation, use the following equation:

$$T_W = \hat{M} - Z_{(1-C)}\left(\sigma / \sqrt{n}\right) - Z_{(1-R)}(\sigma)$$

where

$\hat{M}$ = Mean time to wear-out

σ = Standard deviation

n = Sample size

C = Confidence level

R = Reliability

For $Z_{(1-C)}$ and $Z_{(1-R)}$ use the normal distribution value for a given confidence level and reliability level (see Appendix C, Selected Single-Sided Normal Distribution Probability Points)

Example: A system has a mean time to wear-out of 10,000 hours, a known standard deviation of 600 hours, and a sample size of 16. Determine the time when a component/system should be replaced to avoid wear-out failures so we can be 95% confident that we are 99% reliable.

$$T_W = \hat{M} - Z_{(1-C)}\left(\sigma / \sqrt{n}\right) - Z_{(1-R)}(\sigma)$$

$$T_W = 10,000 - 1.645\left(600 / \sqrt{16}\right) - 2.326(600)$$

$$T_W = 8357.65 \text{ hours}$$

22.2 WEAR-OUT DISTRIBUTION—STANDARD DEVIATION UNKNOWN

If the value of the standard deviation is only an estimate, the confidence of this point estimate is dependent on the sample size, which results in a wider interval. This requires a different approach, and can be calculated using the following equation:

$$T_W = \hat{M} - K_1 s$$

where

$\hat{M}$ = Mean time to wear-out

s = Sample standard deviation

C = Confidence level

K_1 = One-sided factor (see Appendix H, Tolerance Interval Factors)

Example: A system has a mean time to wear-out of 10,000 hours, an estimated standard deviation of 600 hours, and a sample size of 16. Determine the time when a component/system should be replaced to avoid wear-out failures so we can be 95% confident that we are 99% reliable.

$$T_W = \hat{M} - K_1 s = 10,000 - 3.464(600) = 7921.6 \text{ hours}$$

22.2 WEAR-OUT AND CHANCE FAILURE COMBINED

To calculate the probability of not failing due to chance or wear-out, the probability of chance failures is multiplied by the probability of wear-out failures.

For chance failures:

$$R_{(t)} = e^{-\lambda t} \text{ (From Chapter 17)}$$

where

e = The base of natural logarithms 2.718281828 . . . .

λ = Failure rate (recall that $\lambda = 1/\theta$)

t = Time (Note: Time [t] may be substituted for cycles [c])

For wear-out failures:

$$Z = \frac{F_t - T_W}{\sigma \text{ or } s}$$

where

F_t = Failure at time t

T_W = Time to wear-out

σ or s = Standard deviation

For Z use the normal distribution (see Appendix A, Normal Distribution Probability Points—Area below Z)

Probability of lasting to time t:

$$R = Z * R_{(t)}$$

Example: The mean time to failure is 60,000 hours (λ 0.000016667), the confidence limit for the mean was calculated to be 9753.25 hours, and the standard deviation is 600 hours. Calculate the probability of surviving to 8000 hours without a chance failure or wear-out failure.

$$R_{(t)} = e^{-\lambda t} = e^{-0.000016667(8000)} = 0.8752$$

$$Z = \frac{F_t - T_W}{\sigma \text{ or } s} = \frac{8000 - 9753.25}{600} = 2.922 = 0.9982$$

$$R = Z * R_{(t)} = 0.9982 * 0.8752 = 0.8736 \text{ or } 87.36\%$$

23

Conditional Probability of Failure

Sometimes, because of scheduling, a system may be required to operate well into its wear-out phase (note that chance failures may also occur during this period). For example, if a failure occurs, and the impact on the system is not catastrophic, management may want the system to run until a scheduled plant shutdown, which will allow for servicing of the system. If the time of operation to this point is denoted as t_1, the conditional probability can be written as

$$R_{(t|>t_1)} = \frac{R(t)}{R(t_1)}, t > t_1$$

Example: The bearings of an engine have a wear-out distribution that is known to be normal. The distribution has a mean wear-out time of 2000 hours with a standard deviation of 50 hours. The time of operation to this point is 1900 hours. What is the probability that the bearings will survive (wear-out failures only) for the next 16 hours?

$$R_{(t|>t_1)} = \frac{R(t)}{R(t_1)}, t > t_1$$

$$R_{(1916|t>1900)} = \frac{R(1916)}{R(1900)}$$

Note: Z = The normal distribution value (see Appendix A, Normal Distribution Probability Points—Area below Z)

The reliability for 1900 hours is calculated as

$$Z = \left| \frac{1900 - 2000}{50} \right| = 2 = 0.9772$$

The reliability for 1916 hours is calculated as

$$Z = \left| \frac{1916 - 2000}{50} \right| = 1.68 = 0.9535$$

The probability that the system will operate the next 16 hours without a failure is

$$R = \frac{0.9535}{0.9772} = .9757 \text{ or } 97.57\%$$

24
System Reliability

In many cases there are several components that are used in a particular system. When the system comprises many components, and the failure of one component causes a system failure, this is called a *series system*. When redundant components are used, and the failure of one component does not cause a system failure, this is called a *parallel system*.

24.1 SERIES RELIABILITY SYSTEMS

When the system comprises many components, and the failure of one component causes a system failure (see Figure 24.1), the reliability of the system is the product of the component reliability.

The formula to calculate the series reliability system is

$$R_s = (R1)(R2)(R3)...$$

Example: Three components are placed in series, with the following reliabilities: $R1 = 0.95$, $R2 = 0.90$, and $R3 = 0.99$. Calculate the system reliability.

$$R_s = (R1)(R2)(R3) = 0.95 * 0.90 * 0.99 = .8465 \text{ or } 84.65\%$$

24.2 PARALLEL RELIABILITY SYSTEMS

Series systems do not offer a high level of confidence in high-risk situations. Therefore, to increase the confidence and lower risk, redundancy must be used. While redundancy

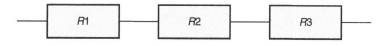

Figure 24.1   Series reliability system.

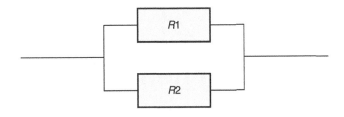

Figure 24.2   Active parallel system.

affords higher confidence in the probability of success of the mission, it also has a penalty of increased cost, weight, space, maintenance, and so on. So trade-offs must be considered.

The most common form of redundancy is *active parallel redundancy*. Active means that both (multiple) components are active and functioning during the life of the system (see Figure 24.2).

Components $R1$ and $R2$ are both active. When the system is active, if both $R1$ and $R2$ are functioning, the system will operate. If $R1$ fails and $R2$ survives, the system will operate. If $R1$ survives and $R2$ fails, the system will operate. When both $R1$ and $R2$ fail, system failure occurs.

For an active parallel system, the reliability can be calculated using the following equation:

$$R_s = 1 - (1 - R1)(1 - R2)...$$

Example: The reliability of $R1$ is 0.95, and the reliability of $R2$ is 0.99. Calculate the system reliability.

$$R_s = 1 - (1 - R1)(1 - R2) = 1 - (1 - 0.95)(1 - 0.99) = 0.9995 \text{ or } 99.95\%$$

24.3 COMBINATION RELIABILITY SYSTEMS

When a system comprises components in parallel and in series (Figure 24.3), reduce the system to a series problem by first solving the parallel reliabilities (Figure 24.4).

Example: Calculate the system reliability for a combination series/parallel system. The reliability of the individual components is as follows:

$R1$	0.95
$R2$	0.90
$R3$	0.99
$R4$	0.95
$R5$	0.85

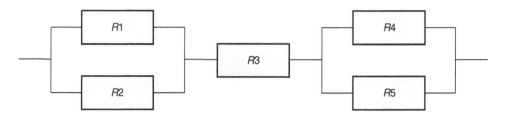

Figure 24.3   Combination series/parallel system.

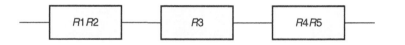

Figure 24.4   Series/parallel system simplified to a series system.
Note: R1 R2 and R4 R5 indicate that the parallel components have been reduced to series components.

Step 1

$$R_1 = 1 - (1 - R1)(1 - R2) = 1 - (1 - 0.95)(1 - 0.90) = 0.9950$$

Step 2

$$R_2 = 1 - (1 - R4)(1 - R5) = 1 - (1 - 0.95)(1 - 0.85) = 0.9925$$

Step 3

$$R_s = (R_1)(R3)(R_3) = 0.9950 * 0.99 * 0.9925 = .9777 \text{ or } 97.77\%$$

25

Stress-Strength Interference

Mechanical systems supporting a known or expected stress have usually been designed with a margin of safety. In situations where risk to human life or strategic mission are concerned, the risk should be kept low while still maintaining economic justification for the project. One method of calculation used for mechanical designs is stress-strength interference. Failure will occur when the stress applied exceeds the strength. In Figure 25.1 the stress and strength do not interfere with each other, so the result is no failure.

In Figure 25.2 a potential for interference between stress and strength (shaded area) exists. When the stress exceeds the strength, failures occur.

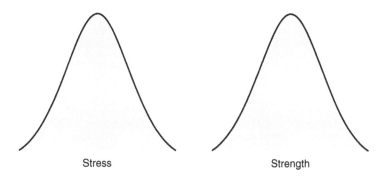

Figure 25.1   Stress and strength without interference.

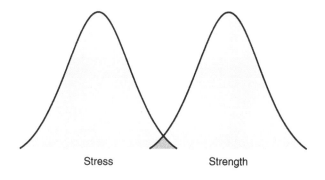

Figure 25.2   Stress and strength with interference.

When the stress-strength distributions are normally distributed, the probability of interference (failure) can be calculated using a Z value with the following equation:

$$Z = \frac{\mu_x - \mu_y}{\sqrt{\sigma_x^2 + \sigma_y^2}}$$

where

μ_x = Average stress

μ_y = Average strength

σ_x = Stress standard deviation

σ_y = Strength standard deviation

For Z use the normal distribution (see Appendix B, Normal Distribution Probability Points—Area above Z).

Example: A shaft has a stress distribution with a mean of 130,000 psi and a standard deviation of 11,000 psi. The shaft has a strength distribution with a mean of 100,000 psi and a standard deviation of 8000 psi. What is the probability of failure?

$$Z = \frac{\mu_x - \mu_y}{\sqrt{\sigma_x^2 + \sigma_y^2}} = \frac{130,000 - 100,000}{\sqrt{11,000^2 + 8,000^2}} = 2.21 = .0136 \text{ or } 1.36\%$$

Appendix A

Normal Distribution Probability Points—Area below Z

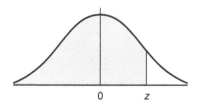

0   z

z	0	0.01	0.02	0.03	0.04	0.05	0.06	0.07	0.08	0.09
0	.5000	.5040	.5080	.5120	.5160	.5199	.5239	.5279	.5319	.5359
0.1	.5398	.5438	.5478	.5517	.5557	.5596	.5636	.5675	.5714	.5753
0.2	.5793	.5832	.5871	.5910	.5948	.5987	.6026	.6064	.6103	.6141
0.3	.6179	.6217	.6255	.6293	.6331	.6368	.6406	.6443	.6480	.6517
0.4	.6554	.6591	.6628	.6664	.6700	.6736	.6772	.6808	.6844	.6879
0.5	.6915	.6950	.6985	.7019	.7054	.7088	.7123	.7157	.7190	.7224
0.6	.7257	.7291	.7324	.7357	.7389	.7422	.7454	.7486	.7517	.7549
0.7	.7580	.7611	.7642	.7673	.7704	.7734	.7764	.7794	.7823	.7852
0.8	.7881	.7910	.7939	.7967	.7995	.8023	.8051	.8078	.8106	.8133
0.9	.8159	.8186	.8212	.8238	.8264	.8289	.8315	.8340	.8365	.8389
1	.8413	.8438	.8461	.8485	.8508	.8531	.8554	.8577	.8599	.8621
1.1	.8643	.8665	.8686	.8708	.8729	.8749	.8770	.8790	.8810	.8830
1.2	.8849	.8869	.8888	.8907	.8925	.8944	.8962	.8980	.8997	.9015
1.3	.9032	.9049	.9066	.9082	.9099	.9115	.9131	.9147	.9162	.9177
1.4	.9192	.9207	.9222	.9236	.9251	.9265	.9279	.9292	.9306	.9319
1.5	.9332	.9345	.9357	.9370	.9382	.9394	.9406	.9418	.9429	.9441
1.6	.9452	.9463	.9474	.9484	.9495	.9505	.9515	.9525	.9535	.9545
1.7	.9554	.9564	.9573	.9582	.9591	.9599	.9608	.9616	.9625	.9633
1.8	.9641	.9649	.9656	.9664	.9671	.9678	.9686	.9693	.9699	.9706
1.9	.9713	.9719	.9726	.9732	.9738	.9744	.9750	.9756	.9761	.9767
2	.9772	.9778	.9783	.9788	.9793	.9798	.9803	.9808	.9812	.9817
2.1	.9821	.9826	.9830	.9834	.9838	.9842	.9846	.9850	.9854	.9857
2.2	.9861	.9864	.9868	.9871	.9875	.9878	.9881	.9884	.9887	.9890
2.3	.9893	.9896	.9898	.9901	.9904	.9906	.9909	.9911	.9913	.9916
2.4	.9918	.9920	.9922	.9925	.9927	.9929	.9931	.9932	.9934	.9936
2.5	.9938	.9940	.9941	.9943	.9945	.9946	.9948	.9949	.9951	.9952
2.6	.9953	.9955	.9956	.9957	.9959	.9960	.9961	.9962	.9963	.9964
2.7	.9965	.9966	.9967	.9968	.9969	.9970	.9971	.9972	.9973	.9974
2.8	.9974	.9975	.9976	.9977	.9977	.9978	.9979	.9979	.9980	.9981
2.9	.9981	.9982	.9982	.9983	.9984	.9984	.9985	.9985	.9986	.9986
3	.9987	.9987	.9987	.9988	.9988	.9989	.9989	.9989	.9990	.9990

Appendix B
Normal Distribution Probability Points—Area above *Z*

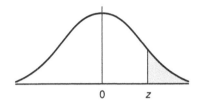

z	0	0.01	0.02	0.03	0.04	0.05	0.06	0.07	0.08	0.09
0	.5000	.4960	.4920	.4880	.4840	.4801	.4761	.4721	.4681	.4641
0.1	.4602	.4562	.4522	.4483	.4443	.4404	.4364	.4325	.4286	.4247
0.2	.4207	.4168	.4129	.4090	.4052	.4013	.3974	.3936	.3897	.3859
0.3	.3821	.3783	.3745	.3707	.3669	.3632	.3594	.3557	.3520	.3483
0.4	.3446	.3409	.3372	.3336	.3300	.3264	.3228	.3192	.3156	.3121
0.5	.3085	.3050	.3015	.2981	.2946	.2912	.2877	.2843	.2810	.2776
0.6	.2743	.2709	.2676	.2643	.2611	.2578	.2546	.2514	.2483	.2451
0.7	.2420	.2389	.2358	.2327	.2296	.2266	.2236	.2206	.2177	.2148
0.8	.2119	.2090	.2061	.2033	.2005	.1977	.1949	.1922	.1894	.1867
0.9	.1841	.1814	.1788	.1762	.1736	.1711	.1685	.1660	.1635	.1611
1	.1587	.1562	.1539	.1515	.1492	.1469	.1446	.1423	.1401	.1379
1.1	.1357	.1335	.1314	.1292	.1271	.1251	.1230	.1210	.1190	.1170
1.2	.1151	.1131	.1112	.1093	.1075	.1056	.1038	.1020	.1003	.0985
1.3	.0968	.0951	.0934	.0918	.0901	.0885	.0869	.0853	.0838	.0823
1.4	.0808	.0793	.0778	.0764	.0749	.0735	.0721	.0708	.0694	.0681
1.5	.0668	.0655	.0643	.0630	.0618	.0606	.0594	.0582	.0571	.0559
1.6	.0548	.0537	.0526	.0516	.0505	.0495	.0485	.0475	.0465	.0455
1.7	.0446	.0436	.0427	.0418	.0409	.0401	.0392	.0384	.0375	.0367
1.8	.0359	.0351	.0344	.0336	.0329	.0322	.0314	.0307	.0301	.0294
1.9	.0287	.0281	.0274	.0268	.0262	.0256	.0250	.0244	.0239	.0233
2	.0228	.0222	.0217	.0212	.0207	.0202	.0197	.0192	.0188	.0183
2.1	.0179	.0174	.0170	.0166	.0162	.0158	.0154	.0150	.0146	.0143
2.2	.0139	.0136	.0132	.0129	.0125	.0122	.0119	.0116	.0113	.0110
2.3	.0107	.0104	.0102	.0099	.0096	.0094	.0091	.0089	.0087	.0084
2.4	.0082	.0080	.0078	.0075	.0073	.0071	.0069	.0068	.0066	.0064
2.5	.0062	.0060	.0059	.0057	.0055	.0054	.0052	.0051	.0049	.0048
2.6	.0047	.0045	.0044	.0043	.0041	.0040	.0039	.0038	.0037	.0036
2.7	.0035	.0034	.0033	.0032	.0031	.0030	.0029	.0028	.0027	.0026
2.8	.0026	.0025	.0024	.0023	.0023	.0022	.0021	.0021	.0020	.0019
2.9	.0019	.0018	.0018	.0017	.0016	.0016	.0015	.0015	.0014	.0014
3	.0013	.0013	.0013	.0012	.0012	.0011	.0011	.0011	.0010	.0010

Appendix C
Selected Single-Sided Normal Distribution Probability Points

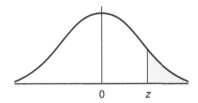

α	z	α	z
0.001	3.090	0.100	1.282
0.005	2.576	0.150	1.036
0.010	2.326	0.200	0.842
0.015	2.170	0.300	0.524
0.020	2.054	0.400	0.253
0.025	1.960	0.500	0.000
0.050	1.645	0.600	−0.253

Appendix D
Selected Double-Sided Normal Distribution Probability Points

*Do not divide by 2; the table has already compensated for this.

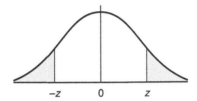

α	z	α	z
0.001	3.291	0.100	1.645
0.005	2.807	0.150	1.440
0.010	2.576	0.200	1.282
0.015	2.432	0.300	1.036
0.020	2.326	0.400	0.842
0.025	2.241	0.500	0.675
0.050	1.960	0.600	0.524

Appendix E

Percentage Points of the Student's *t*-Distribution

df	α	p 0.9 / 0.1	0.95 / 0.05	0.975 / 0.025	0.9875 / 0.0125	0.995 / 0.005	0.9975 / 0.0025	0.999 / 0.001	0.9995 / 0.0005
1		3.078	6.314	12.706	25.452	63.657	127.321	318.309	636.619
2		1.886	2.920	4.303	6.205	9.925	14.089	22.327	31.599
3		1.638	2.353	3.182	4.177	5.841	7.453	10.215	12.924
4		1.533	2.132	2.776	3.495	4.604	5.598	7.173	8.610
5		1.476	2.015	2.571	3.163	4.032	4.773	5.893	6.869
6		1.440	1.943	2.447	2.969	3.707	4.317	5.208	5.959
7		1.415	1.895	2.365	2.841	3.499	4.029	4.785	5.408
8		1.397	1.860	2.306	2.752	3.355	3.833	4.501	5.041
9		1.383	1.833	2.262	2.685	3.250	3.690	4.297	4.781
10		1.372	1.812	2.228	2.634	3.169	3.581	4.144	4.587
11		1.363	1.796	2.201	2.593	3.106	3.497	4.025	4.437
12		1.356	1.782	2.179	2.560	3.055	3.428	3.930	4.318
13		1.350	1.771	2.160	2.533	3.012	3.372	3.852	4.221
14		1.345	1.761	2.145	2.510	2.977	3.326	3.787	4.140
15		1.341	1.753	2.131	2.490	2.947	3.286	3.733	4.073
16		1.337	1.746	2.120	2.473	2.921	3.252	3.686	4.015
17		1.333	1.740	2.110	2.458	2.898	3.222	3.646	3.965
18		1.330	1.734	2.101	2.445	2.878	3.197	3.610	3.922
19		1.328	1.729	2.093	2.433	2.861	3.174	3.579	3.883
20		1.325	1.725	2.086	2.423	2.845	3.153	3.552	3.850
21		1.323	1.721	2.080	2.414	2.831	3.135	3.527	3.819

Continued

Continued

df	α	p 0.9	0.95	0.975	0.9875	0.995	0.9975	0.999	0.9995
		0.1	0.05	0.025	0.0125	0.005	0.0025	0.001	0.0005
22		1.321	1.717	2.074	2.405	2.819	3.119	3.505	3.792
23		1.319	1.714	2.069	2.398	2.807	3.104	3.485	3.768
24		1.318	1.711	2.064	2.391	2.797	3.090	3.467	3.745
25		1.316	1.708	2.060	2.385	2.787	3.078	3.450	3.725
26		1.315	1.706	2.056	2.379	2.779	3.067	3.435	3.707
27		1.314	1.703	2.052	2.373	2.771	3.057	3.421	3.690
28		1.313	1.701	2.048	2.368	2.763	3.047	3.408	3.674
29		1.311	1.699	2.045	2.364	2.756	3.038	3.396	3.659
30		1.310	1.697	2.042	2.360	2.750	3.030	3.385	3.646
60		1.296	1.671	2.000	2.299	2.660	2.915	3.232	3.460
90		1.291	1.662	1.987	2.280	2.632	2.878	3.183	3.402
120		1.289	1.658	1.980	2.270	2.617	2.860	3.160	3.373
∞		1.282	1.645	1.960	2.242	2.576	2.807	3.090	3.291

Appendix F

Distribution of the Chi-Square

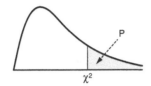

df	0.995	0.99	0.95	0.9	0.1	0.05	0.01	0.005
1	—	—	0.004	0.016	2.706	3.841	6.635	7.879
2	0.010	0.020	0.103	0.211	4.605	5.991	9.210	10.597
3	0.072	0.115	0.352	0.584	6.251	7.815	11.345	12.838
4	0.207	0.297	0.711	1.064	7.779	9.488	13.277	14.860
5	0.412	0.554	1.145	1.610	9.236	11.070	15.086	16.750
6	0.676	0.872	1.635	2.204	10.645	12.592	16.812	18.548
7	0.989	1.239	2.167	2.833	12.017	14.067	18.475	20.278
8	1.344	1.646	2.733	3.490	13.362	15.507	20.090	21.955
9	1.735	2.088	3.325	4.168	14.684	16.919	21.666	23.589
10	2.156	2.558	3.940	4.865	15.987	18.307	23.209	25.188
11	2.603	3.053	4.575	5.578	17.275	19.675	24.725	26.757
12	3.074	3.571	5.226	6.304	18.549	21.026	26.217	28.300
13	3.565	4.107	5.892	7.042	19.812	22.362	27.688	29.819
14	4.075	4.660	6.571	7.790	21.064	23.685	29.141	31.319
15	4.601	5.229	7.261	8.547	22.307	24.996	30.578	32.801
16	5.142	5.812	7.962	9.312	23.542	26.296	32.000	34.267
17	5.697	6.408	8.672	10.085	24.769	27.587	33.409	35.718
18	6.265	7.015	9.390	10.865	25.989	28.869	34.805	37.156
19	6.844	7.633	10.117	11.651	27.204	30.144	36.191	38.582
20	7.434	8.260	10.851	12.443	28.412	31.410	37.566	39.997
21	8.034	8.897	11.591	13.240	29.615	32.671	38.932	41.401
22	8.643	9.542	12.338	14.041	30.813	33.924	40.289	42.796

Continued

Continued

df	0.995	0.99	0.95	0.9	0.1	0.05	0.01	0.005
23	9.260	10.196	13.091	14.848	32.007	35.172	41.638	44.181
24	9.886	10.856	13.848	15.659	33.196	36.415	42.980	45.559
25	10.520	11.524	14.611	16.473	34.382	37.652	44.314	46.928
26	11.160	12.198	15.379	17.292	35.563	38.885	45.642	48.290
27	11.808	12.879	16.151	18.114	36.741	40.113	46.963	49.645
28	12.461	13.565	16.928	18.939	37.916	41.337	48.278	50.993
29	13.121	14.256	17.708	19.768	39.087	42.557	49.588	52.336
30	13.787	14.953	18.493	20.599	40.256	43.773	50.892	53.672
50	27.99	29.71	34.76	37.69	63.167	67.505	76.154	79.49
100	67.328	70.065	77.929	82.358	118.498	124.342	135.807	140.169

Appendix G
Percentages of the *F*-Distribution

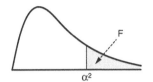

Percentages of the *F*-Distribution α = 0.01

	Degrees of freedom for numerator v1									
v2	1	2	3	4	5	6	7	8	9	10
1	4052	4999	5404	5624	5764	5859	5928	5981	6022	6056
2	98.50	99.00	99.20	99.30	99.30	99.30	99.40	99.40	99.40	99.40
3	34.10	30.80	29.50	28.70	28.20	27.90	27.70	27.50	27.30	27.20
4	21.20	18.00	16.70	16.00	15.50	15.20	15.00	14.80	14.70	14.50
5	16.30	13.30	12.10	11.40	11.00	10.70	10.50	10.30	10.20	10.10
6	13.70	10.90	9.78	9.15	8.75	8.47	8.26	8.10	7.98	7.87
7	12.20	9.55	8.45	7.85	7.46	7.19	6.99	6.84	6.72	6.62
8	11.30	8.65	7.59	7.01	6.63	6.37	6.18	6.03	5.91	5.81
9	10.60	8.02	6.99	6.42	6.06	5.80	5.61	5.47	5.35	5.26
10	10.00	7.56	6.55	5.99	5.64	5.39	5.20	5.06	4.94	4.85
11	9.65	7.21	6.22	5.67	5.32	5.07	4.89	4.74	4.63	4.54
12	9.33	6.93	5.95	5.41	5.06	4.82	4.64	4.50	4.39	4.30
13	9.07	6.70	5.74	5.21	4.86	4.62	4.44	4.30	4.19	4.10
14	8.86	6.51	5.56	5.04	4.69	4.46	4.28	4.14	4.03	3.94
15	8.68	6.36	5.42	4.89	4.56	4.32	4.14	4.00	3.89	3.80
16	8.53	6.23	5.29	4.77	4.44	4.20	4.03	3.89	3.78	3.69
17	8.40	6.11	5.19	4.67	4.34	4.10	3.93	3.79	3.68	3.59
18	8.29	6.01	5.09	4.58	4.25	4.01	3.84	3.71	3.60	3.51
19	8.18	5.93	5.01	4.50	4.17	3.94	3.77	3.63	3.52	3.43
20	8.10	5.85	4.94	4.43	4.10	3.87	3.70	3.56	3.46	3.37
21	8.02	5.78	4.87	4.37	4.04	3.81	3.64	3.51	3.40	3.31
22	7.95	5.72	4.82	4.31	3.99	3.76	3.59	3.45	3.35	3.26
23	7.88	5.66	4.76	4.26	3.94	3.71	3.54	3.41	3.30	3.21
24	7.82	5.61	4.72	4.22	3.90	3.67	3.50	3.36	3.26	3.17
25	7.77	5.57	4.68	4.18	3.85	3.63	3.46	3.32	3.22	3.13
26	7.72	5.53	4.64	4.14	3.82	3.59	3.42	3.29	3.18	3.09
27	7.68	5.49	4.60	4.11	3.78	3.56	3.39	3.26	3.15	3.06
28	7.64	5.45	4.57	4.07	3.75	3.53	3.36	3.23	3.12	3.03
29	7.60	5.42	4.54	4.04	3.73	3.50	3.33	3.20	3.09	3.00
30	7.56	5.39	4.51	4.02	3.70	3.47	3.30	3.17	3.07	2.98
40	7.31	5.18	4.31	3.83	3.51	3.29	3.12	2.99	2.89	2.80
60	7.08	4.98	4.13	3.65	3.34	3.12	2.95	2.82	2.72	2.63
120	6.85	4.79	3.95	3.48	3.17	2.96	2.79	2.66	2.56	2.47
∞	6.64	4.61	3.78	3.32	3.02	2.80	2.64	2.51	2.41	2.32

Continued

	Percentages of the *F*-Distribution α = 0.01								
	Degrees of freedom for numerator ν1								
ν2	12	15	20	24	30	40	60	120	∞
1	6107	6157	6209	6234	6260	6286	6313	6340	6366
2	99.40	99.40	99.40	99.50	99.50	99.50	99.50	99.50	99.50
3	27.10	26.90	26.70	26.60	26.50	26.40	26.30	26.20	26.10
4	14.40	14.20	14.00	13.90	13.80	13.70	13.70	13.60	13.50
5	9.89	9.72	9.55	9.47	9.38	9.29	9.20	9.11	9.02
6	7.72	7.56	7.40	7.31	7.23	7.14	7.06	6.97	6.88
7	6.47	6.31	6.16	6.07	5.99	5.91	5.82	5.74	5.65
8	5.67	5.52	5.36	5.28	5.20	5.12	5.03	4.95	4.86
9	5.11	4.96	4.81	4.73	4.65	4.57	4.48	4.40	4.31
10	4.71	4.56	4.41	4.33	4.25	4.17	4.08	4.00	3.91
11	4.40	4.25	4.10	4.02	3.94	3.86	3.78	3.69	3.60
12	4.16	4.01	3.86	3.78	3.70	3.62	3.54	3.45	3.36
13	3.96	3.82	3.66	3.59	3.51	3.43	3.34	3.25	3.17
14	3.80	3.66	3.51	3.43	3.35	3.27	3.18	3.09	3.00
15	3.67	3.52	3.37	3.29	3.21	3.13	3.05	2.96	2.87
16	3.55	3.41	3.26	3.18	3.10	3.02	2.93	2.84	2.75
17	3.46	3.31	3.16	3.08	3.00	2.92	2.83	2.75	2.65
18	3.37	3.23	3.08	3.00	2.92	2.84	2.75	2.66	2.57
19	3.30	3.15	3.00	2.92	2.84	2.76	2.67	2.58	2.49
20	3.23	3.09	2.94	2.86	2.78	2.69	2.61	2.52	2.42
21	3.17	3.03	2.88	2.80	2.72	2.64	2.55	2.46	2.36
22	3.12	2.98	2.83	2.75	2.67	2.58	2.50	2.40	2.31
23	3.07	2.93	2.78	2.70	2.62	2.54	2.45	2.35	2.26
24	3.03	2.89	2.74	2.66	2.58	2.49	2.40	2.31	2.21
25	2.99	2.85	2.70	2.62	2.54	2.45	2.36	2.27	2.17
26	2.96	2.81	2.66	2.58	2.50	2.42	2.33	2.23	2.13
27	2.93	2.78	2.63	2.55	2.47	2.38	2.29	2.20	2.10
28	2.90	2.75	2.60	2.52	2.44	2.35	2.26	2.17	2.06
29	2.87	2.73	2.57	2.49	2.41	2.33	2.23	2.14	2.03
30	2.84	2.70	2.55	2.47	2.39	2.30	2.21	2.11	2.01
40	2.66	2.52	2.37	2.29	2.20	2.11	2.02	1.92	1.80
60	2.50	2.35	2.20	2.12	2.03	1.94	1.84	1.73	1.60
120	2.34	2.19	2.03	1.95	1.86	1.76	1.66	1.53	1.38
∞	2.18	2.04	1.88	1.79	1.70	1.59	1.47	1.32	1.00

	Percentages of the *F*-Distribution α = 0.025									
	Degrees of freedom for numerator ν1									
ν2	1	2	3	4	5	6	7	8	9	10
1	648	799	864	900	922	937	948	957	963	969
2	38.50	39.00	39.20	39.20	39.30	39.30	39.40	39.40	39.40	39.40
3	17.40	16.00	15.40	15.10	14.90	14.70	14.60	14.50	14.50	14.40
4	12.20	10.60	9.98	9.60	9.36	9.20	9.07	8.98	8.90	8.84
5	10.00	8.43	7.76	7.39	7.15	6.98	6.85	6.76	6.68	6.62
6	8.81	7.26	6.60	6.23	5.99	5.82	5.70	5.60	5.52	5.46
7	8.07	6.54	5.89	5.52	5.29	5.12	4.99	4.90	4.82	4.76
8	7.57	6.06	5.42	5.05	4.82	4.65	4.53	4.43	4.36	4.30
9	7.21	5.71	5.08	4.72	4.48	4.32	4.20	4.10	4.03	3.96
10	6.94	5.46	4.83	4.47	4.24	4.07	3.95	3.85	3.78	3.72
11	6.72	5.26	4.63	4.28	4.04	3.88	3.76	3.66	3.59	3.53
12	6.55	5.10	4.47	4.12	3.89	3.73	3.61	3.51	3.44	3.37
13	6.41	4.97	4.35	4.00	3.77	3.60	3.48	3.39	3.31	3.25
14	6.30	4.86	4.24	3.89	3.66	3.50	3.38	3.29	3.21	3.15
15	6.20	4.77	4.15	3.80	3.58	3.41	3.29	3.20	3.12	3.06
16	6.12	4.69	4.08	3.73	3.50	3.34	3.22	3.12	3.05	2.99
17	6.04	4.62	4.01	3.66	3.44	3.28	3.16	3.06	2.98	2.92
18	5.98	4.56	3.95	3.61	3.38	3.22	3.10	3.01	2.93	2.87
19	5.92	4.51	3.90	3.56	3.33	3.17	3.05	2.96	2.88	2.82
20	5.87	4.46	3.86	3.51	3.29	3.13	3.01	2.91	2.84	2.77
21	5.83	4.42	3.82	3.48	3.25	3.09	2.97	2.87	2.80	2.73
22	5.79	4.38	3.78	3.44	3.22	3.05	2.93	2.84	2.76	2.70
23	5.75	4.35	3.75	3.41	3.18	3.02	2.90	2.81	2.73	2.67
24	5.72	4.32	3.72	3.38	3.15	2.99	2.87	2.78	2.70	2.64
25	5.69	4.29	3.69	3.35	3.13	2.97	2.85	2.75	2.68	2.61
26	5.66	4.27	3.67	3.33	3.10	2.94	2.82	2.73	2.65	2.59
27	5.63	4.24	3.65	3.31	3.08	2.92	2.80	2.71	2.63	2.57
28	5.61	4.22	3.63	3.29	3.06	2.90	2.78	2.69	2.61	2.55
29	5.59	4.20	3.61	3.27	3.04	2.88	2.76	2.67	2.59	2.53
30	5.57	4.18	3.59	3.25	3.03	2.87	2.75	2.65	2.57	2.51
40	5.42	4.05	3.46	3.13	2.90	2.74	2.62	2.53	2.45	2.39
60	5.29	3.93	3.34	3.01	2.79	2.63	2.51	2.41	2.33	2.27
120	5.15	3.80	3.23	2.89	2.67	2.52	2.39	2.30	2.22	2.16
∞	5.02	3.69	3.12	2.79	2.57	2.41	2.29	2.19	2.11	2.05

Continued

	Percentages of the *F*-Distribution α = 0.025								
	Degrees of freedom for numerator ν1								
ν2	12	15	20	24	30	40	60	120	∞
1	977	985	993	997	1001	1006	1010	1014	1018
2	39.40	39.40	39.40	39.50	39.50	39.50	39.50	39.50	39.50
3	14.30	14.30	14.20	14.10	14.10	14.00	14.00	13.90	13.90
4	8.75	8.66	8.56	8.51	8.46	8.41	8.36	8.31	8.26
5	6.52	6.43	6.33	6.28	6.23	6.18	6.12	6.07	6.02
6	5.37	5.27	5.17	5.12	5.07	5.01	4.96	4.90	4.85
7	4.67	4.57	4.47	4.41	4.36	4.31	4.25	4.20	4.14
8	4.20	4.10	4.00	3.95	3.89	3.84	3.78	3.73	3.67
9	3.87	3.77	3.67	3.61	3.56	3.51	3.45	3.39	3.33
10	3.62	3.52	3.42	3.37	3.31	3.26	3.20	3.14	3.08
11	3.43	3.33	3.23	3.17	3.12	3.06	3.00	2.94	2.88
12	3.28	3.18	3.07	3.02	2.96	2.91	2.85	2.79	2.73
13	3.15	3.05	2.95	2.89	2.84	2.78	2.72	2.66	2.60
14	3.05	2.95	2.84	2.79	2.73	2.67	2.61	2.55	2.49
15	2.96	2.86	2.76	2.70	2.64	2.59	2.52	2.46	2.40
16	2.89	2.79	2.68	2.63	2.57	2.51	2.45	2.38	2.32
17	2.82	2.72	2.62	2.56	2.50	2.44	2.38	2.32	2.25
18	2.77	2.67	2.56	2.50	2.44	2.38	2.32	2.26	2.19
19	2.72	2.62	2.51	2.45	2.39	2.33	2.27	2.20	2.13
20	2.68	2.57	2.46	2.41	2.35	2.29	2.22	2.16	2.09
21	2.64	2.53	2.42	2.37	2.31	2.25	2.18	2.11	2.04
22	2.60	2.50	2.39	2.33	2.27	2.21	2.14	2.08	2.00
23	2.57	2.47	2.36	2.30	2.24	2.18	2.11	2.04	1.97
24	2.54	2.44	2.33	2.27	2.21	2.15	2.08	2.01	1.94
25	2.51	2.41	2.30	2.24	2.18	2.12	2.05	1.98	1.91
26	2.49	2.39	2.28	2.22	2.16	2.09	2.03	1.95	1.88
27	2.47	2.36	2.25	2.19	2.13	2.07	2.00	1.93	1.85
28	2.45	2.34	2.23	2.17	2.11	2.05	1.98	1.91	1.83
29	2.43	2.32	2.21	2.15	2.09	2.03	1.96	1.89	1.81
30	2.41	2.31	2.20	2.14	2.07	2.01	1.94	1.87	1.79
40	2.29	2.18	2.07	2.01	1.94	1.88	1.80	1.72	1.64
60	2.17	2.06	1.94	1.88	1.82	1.74	1.67	1.58	1.48
120	2.05	1.94	1.82	1.76	1.69	1.61	1.53	1.43	1.31
∞	1.94	1.83	1.71	1.64	1.57	1.48	1.39	1.27	1.00

	Percentages of the *F*-Distribution α = 0.05									
	Degrees of freedom for numerator ν1									
ν2	1	2	3	4	5	6	7	8	9	10
1	161	199	216	225	230	234	237	239	241	242
2	18.50	19.00	19.20	19.20	19.30	19.30	19.40	19.40	19.40	19.40
3	10.10	9.55	9.28	9.12	9.01	8.94	8.89	8.85	8.81	8.79
4	7.71	6.94	6.59	6.39	6.26	6.16	6.09	6.04	6.00	5.96
5	6.61	5.79	5.41	5.19	5.05	4.95	4.88	4.82	4.77	4.74
6	5.99	5.14	4.76	4.53	4.39	4.28	4.21	4.15	4.10	4.06
7	5.59	4.74	4.35	4.12	3.97	3.87	3.79	3.73	3.68	3.64
8	5.32	4.46	4.07	3.84	3.69	3.58	3.50	3.44	3.39	3.35
9	5.12	4.26	3.86	3.63	3.48	3.37	3.29	3.23	3.18	3.14
10	4.96	4.10	3.71	3.48	3.33	3.22	3.14	3.07	3.02	2.98
11	4.84	3.98	3.59	3.36	3.20	3.09	3.01	2.95	2.90	2.85
12	4.75	3.89	3.49	3.26	3.11	3.00	2.91	2.85	2.80	2.75
13	4.67	3.81	3.41	3.18	3.03	2.92	2.83	2.77	2.71	2.67
14	4.60	3.74	3.34	3.11	2.96	2.85	2.76	2.70	2.65	2.60
15	4.54	3.68	3.29	3.06	2.90	2.79	2.71	2.64	2.59	2.54
16	4.49	3.63	3.24	3.01	2.85	2.74	2.66	2.59	2.54	2.49
17	4.45	3.59	3.20	2.96	2.81	2.70	2.61	2.55	2.49	2.45
18	4.41	3.55	3.16	2.93	2.77	2.66	2.58	2.51	2.46	2.41
19	4.38	3.52	3.13	2.90	2.74	2.63	2.54	2.48	2.42	2.38
20	4.35	3.49	3.10	2.87	2.71	2.60	2.51	2.45	2.39	2.35
21	4.32	3.47	3.07	2.84	2.68	2.57	2.49	2.42	2.37	2.32
22	4.30	3.44	3.05	2.82	2.66	2.55	2.46	2.40	2.34	2.30
23	4.28	3.42	3.03	2.80	2.64	2.53	2.44	2.37	2.32	2.27
24	4.26	3.40	3.01	2.78	2.62	2.51	2.42	2.36	2.30	2.25
25	4.24	3.39	2.99	2.76	2.60	2.49	2.40	2.34	2.28	2.24
26	4.23	3.37	2.98	2.74	2.59	2.47	2.39	2.32	2.27	2.22
27	4.21	3.35	2.96	2.73	2.57	2.46	2.37	2.31	2.25	2.20
28	4.20	3.34	2.95	2.71	2.56	2.45	2.36	2.29	2.24	2.19
29	4.18	3.33	2.93	2.70	2.55	2.43	2.35	2.28	2.22	2.18
30	4.17	3.32	2.92	2.69	2.53	2.42	2.33	2.27	2.21	2.16
40	4.08	3.23	2.84	2.61	2.45	2.34	2.25	2.18	2.12	2.08
60	4.00	3.15	2.76	2.53	2.37	2.25	2.17	2.10	2.04	1.99
120	3.92	3.07	2.68	2.45	2.29	2.18	2.09	2.02	1.96	1.91
∞	3.84	3.00	2.60	2.37	2.21	2.10	2.01	1.94	1.88	1.83

Continued

	Percentages of the *F*-Distribution α = 0.05								
	Degrees of freedom for numerator ν1								
ν2	12	15	20	24	30	40	60	120	∞
1	244	246	248	249	250	251	252	253	254
2	19.40	19.40	19.40	19.50	19.50	19.50	19.50	19.50	19.50
3	8.74	8.70	8.66	8.64	8.62	8.59	8.57	8.55	8.53
4	5.91	5.86	5.80	5.77	5.75	5.72	5.69	5.66	5.63
5	4.68	4.62	4.56	4.53	4.50	4.46	4.43	4.40	4.37
6	4.00	3.94	3.87	3.84	3.81	3.77	3.74	3.70	3.67
7	3.57	3.51	3.44	3.41	3.38	3.34	3.30	3.27	3.23
8	3.28	3.22	3.15	3.12	3.08	3.04	3.01	2.97	2.93
9	3.07	3.01	2.94	2.90	2.86	2.83	2.79	2.75	2.71
10	2.91	2.85	2.77	2.74	2.70	2.66	2.62	2.58	2.54
11	2.79	2.72	2.65	2.61	2.57	2.53	2.49	2.45	2.40
12	2.69	2.62	2.54	2.51	2.47	2.43	2.38	2.34	2.30
13	2.60	2.53	2.46	2.42	2.38	2.34	2.30	2.25	2.21
14	2.53	2.46	2.39	2.35	2.31	2.27	2.22	2.18	2.13
15	2.48	2.40	2.33	2.29	2.25	2.20	2.16	2.11	2.07
16	2.42	2.35	2.28	2.24	2.19	2.15	2.11	2.06	2.01
17	2.38	2.31	2.23	2.19	2.15	2.10	2.06	2.01	1.96
18	2.34	2.27	2.19	2.15	2.11	2.06	2.02	1.97	1.92
19	2.31	2.23	2.16	2.11	2.07	2.03	1.98	1.93	1.88
20	2.28	2.20	2.12	2.08	2.04	1.99	1.95	1.90	1.84
21	2.25	2.18	2.10	2.05	2.01	1.96	1.92	1.87	1.81
22	2.23	2.15	2.07	2.03	1.98	1.94	1.89	1.84	1.78
23	2.20	2.13	2.05	2.01	1.96	1.91	1.86	1.81	1.76
24	2.18	2.11	2.03	1.98	1.94	1.89	1.84	1.79	1.73
25	2.16	2.09	2.01	1.96	1.92	1.87	1.82	1.77	1.71
26	2.15	2.07	1.99	1.95	1.90	1.85	1.80	1.75	1.69
27	2.13	2.06	1.97	1.93	1.88	1.84	1.79	1.73	1.67
28	2.12	2.04	1.96	1.91	1.87	1.82	1.77	1.71	1.65
29	2.10	2.03	1.94	1.90	1.85	1.81	1.75	1.70	1.64
30	2.09	2.01	1.93	1.89	1.84	1.79	1.74	1.68	1.62
40	2.00	1.92	1.84	1.79	1.74	1.69	1.64	1.58	1.51
60	1.92	1.84	1.75	1.70	1.65	1.59	1.53	1.47	1.39
120	1.83	1.75	1.66	1.61	1.55	1.50	1.43	1.35	1.25
∞	1.75	1.67	1.57	1.52	1.46	1.39	1.32	1.22	1.00

	Percentages of the *F*-Distribution α = 0.10									
	Degrees of freedom for numerator ν1									
ν2	1	2	3	4	5	6	7	8	9	10
1	40	50	54	56	57	58	59	59	60	60
2	8.53	9.00	9.16	9.24	9.29	9.33	9.35	9.37	9.38	9.39
3	5.54	5.46	5.39	5.34	5.31	5.28	5.27	5.25	5.24	5.23
4	4.54	4.32	4.19	4.11	4.05	4.01	3.98	3.95	3.94	3.92
5	4.06	3.78	3.62	3.52	3.45	3.40	3.37	3.34	3.32	3.30
6	3.78	3.46	3.29	3.18	3.11	3.05	3.01	2.98	2.96	2.94
7	3.59	3.26	3.07	2.96	2.88	2.83	2.78	2.75	2.72	2.70
8	3.46	3.11	2.92	2.81	2.73	2.67	2.62	2.59	2.56	2.54
9	3.36	3.01	2.81	2.69	2.61	2.55	2.51	2.47	2.44	2.42
10	3.29	2.92	2.73	2.61	2.52	2.46	2.41	2.38	2.35	2.32
11	3.23	2.86	2.66	2.54	2.45	2.39	2.34	2.30	2.27	2.25
12	3.18	2.81	2.61	2.48	2.39	2.33	2.28	2.24	2.21	2.19
13	3.14	2.76	2.56	2.43	2.35	2.28	2.23	2.20	2.16	2.14
14	3.10	2.73	2.52	2.39	2.31	2.24	2.19	2.15	2.12	2.10
15	3.07	2.70	2.49	2.36	2.27	2.21	2.16	2.12	2.09	2.06
16	3.05	2.67	2.46	2.33	2.24	2.18	2.13	2.09	2.06	2.03
17	3.03	2.64	2.44	2.31	2.22	2.15	2.10	2.06	2.03	2.00
18	3.01	2.62	2.42	2.29	2.20	2.13	2.08	2.04	2.00	1.98
19	2.99	2.61	2.40	2.27	2.18	2.11	2.06	2.02	1.98	1.96
20	2.97	2.59	2.38	2.25	2.16	2.09	2.04	2.00	1.96	1.94
21	2.96	2.57	2.36	2.23	2.14	2.08	2.02	1.98	1.95	1.92
22	2.95	2.56	2.35	2.22	2.13	2.06	2.01	1.97	1.93	1.90
23	2.94	2.55	2.34	2.21	2.11	2.05	1.99	1.95	1.92	1.89
24	2.93	2.54	2.33	2.19	2.10	2.04	1.98	1.94	1.91	1.88
25	2.92	2.53	2.32	2.18	2.09	2.02	1.97	1.93	1.89	1.87
26	2.91	2.52	2.31	2.17	2.08	2.01	1.96	1.92	1.88	1.86
27	2.90	2.51	2.30	2.17	2.07	2.00	1.95	1.91	1.87	1.85
28	2.89	2.50	2.29	2.16	2.06	2.00	1.94	1.90	1.87	1.84
29	2.89	2.50	2.28	2.15	2.06	1.99	1.93	1.89	1.86	1.83
30	2.88	2.49	2.28	2.14	2.05	1.98	1.93	1.88	1.85	1.82
40	2.84	2.44	2.23	2.09	2.00	1.93	1.87	1.83	1.79	1.76
60	2.79	2.39	2.18	2.04	1.95	1.87	1.82	1.77	1.74	1.71
120	2.75	2.35	2.13	1.99	1.90	1.82	1.77	1.72	1.68	1.65
∞	2.71	2.30	2.08	1.94	1.85	1.77	1.72	1.67	1.63	1.60

Continued

Percentages of the *F*-Distribution α = 0.10

	Degrees of freedom for numerator ν1								
ν2	12	15	20	24	30	40	60	120	∞
1	61	61	62	62	62	63	63	63	63
2	9.41	9.42	9.44	9.45	9.46	9.47	9.47	9.48	9.49
3	5.22	5.20	5.18	5.18	5.17	5.16	5.15	5.14	5.13
4	3.90	3.87	3.84	3.83	3.82	3.80	3.79	3.78	3.76
5	3.27	3.24	3.21	3.19	3.17	3.16	3.14	3.12	3.11
6	2.90	2.87	2.84	2.82	2.80	2.78	2.76	2.74	2.72
7	2.67	2.63	2.59	2.58	2.56	2.54	2.51	2.49	2.47
8	2.50	2.46	2.42	2.40	2.38	2.36	2.34	2.32	2.29
9	2.38	2.34	2.30	2.28	2.25	2.23	2.21	2.18	2.16
10	2.28	2.24	2.20	2.18	2.16	2.13	2.11	2.08	2.06
11	2.21	2.17	2.12	2.10	2.08	2.05	2.03	2.00	1.97
12	2.15	2.10	2.06	2.04	2.01	1.99	1.96	1.93	1.90
13	2.10	2.05	2.01	1.98	1.96	1.93	1.90	1.88	1.85
14	2.05	2.01	1.96	1.94	1.91	1.89	1.86	1.83	1.80
15	2.02	1.97	1.92	1.90	1.87	1.85	1.82	1.79	1.76
16	1.99	1.94	1.89	1.87	1.84	1.81	1.78	1.75	1.72
17	1.96	1.91	1.86	1.84	1.81	1.78	1.75	1.72	1.69
18	1.93	1.89	1.84	1.81	1.78	1.75	1.72	1.69	1.66
19	1.91	1.86	1.81	1.79	1.76	1.73	1.70	1.67	1.63
20	1.89	1.84	1.79	1.77	1.74	1.71	1.68	1.64	1.61
21	1.87	1.83	1.78	1.75	1.72	1.69	1.66	1.62	1.59
22	1.86	1.81	1.76	1.73	1.70	1.67	1.64	1.60	1.57
23	1.84	1.80	1.74	1.72	1.69	1.66	1.62	1.59	1.55
24	1.83	1.78	1.73	1.70	1.67	1.64	1.61	1.57	1.53
25	1.82	1.77	1.72	1.69	1.66	1.63	1.59	1.56	1.52
26	1.81	1.76	1.71	1.68	1.65	1.61	1.58	1.54	1.50
27	1.80	1.75	1.70	1.67	1.64	1.60	1.57	1.53	1.49
28	1.79	1.74	1.69	1.66	1.63	1.59	1.56	1.52	1.48
29	1.78	1.73	1.68	1.65	1.62	1.58	1.55	1.51	1.47
30	1.77	1.72	1.67	1.64	1.61	1.57	1.54	1.50	1.46
40	1.71	1.66	1.61	1.57	1.54	1.51	1.47	1.42	1.38
60	1.66	1.60	1.54	1.51	1.48	1.44	1.40	1.35	1.29
120	1.60	1.55	1.48	1.45	1.41	1.37	1.32	1.26	1.19
∞	1.55	1.49	1.42	1.38	1.34	1.30	1.24	1.17	1.00

	Percentages of the *F*-Distribution α = 0.25									
	Degrees of freedom for numerator ν1									
ν2	1	2	3	4	5	6	7	8	9	10
1	6	8	8	9	9	9	9	9	9	9
2	2.57	3.00	3.15	3.23	3.28	3.31	3.34	3.35	3.37	3.38
3	2.02	2.28	2.36	2.39	2.41	2.42	2.43	2.44	2.44	2.44
4	1.81	2.00	2.05	2.06	2.07	2.08	2.08	2.08	2.08	2.08
5	1.69	1.85	1.88	1.89	1.89	1.89	1.89	1.89	1.89	1.89
6	1.62	1.76	1.78	1.79	1.79	1.78	1.78	1.78	1.77	1.77
7	1.57	1.70	1.72	1.72	1.71	1.71	1.70	1.70	1.69	1.69
8	1.54	1.66	1.67	1.66	1.66	1.65	1.64	1.64	1.63	1.63
9	1.51	1.62	1.63	1.63	1.62	1.61	1.60	1.60	1.59	1.59
10	1.49	1.60	1.60	1.59	1.59	1.58	1.57	1.56	1.56	1.55
11	1.47	1.58	1.58	1.57	1.56	1.55	1.54	1.53	1.53	1.52
12	1.46	1.56	1.56	1.55	1.54	1.53	1.52	1.51	1.51	1.50
13	1.45	1.55	1.55	1.53	1.52	1.51	1.50	1.49	1.49	1.48
14	1.44	1.53	1.53	1.52	1.51	1.50	1.49	1.48	1.47	1.46
15	1.43	1.52	1.52	1.51	1.49	1.48	1.47	1.46	1.46	1.45
16	1.42	1.51	1.51	1.50	1.48	1.47	1.46	1.45	1.44	1.44
17	1.42	1.51	1.50	1.49	1.47	1.46	1.45	1.44	1.43	1.43
18	1.41	1.50	1.49	1.48	1.46	1.45	1.44	1.43	1.42	1.42
19	1.41	1.49	1.49	1.47	1.46	1.44	1.43	1.42	1.41	1.41
20	1.40	1.49	1.48	1.47	1.45	1.44	1.43	1.42	1.41	1.40
21	1.40	1.48	1.48	1.46	1.44	1.43	1.42	1.41	1.40	1.39
22	1.40	1.48	1.47	1.45	1.44	1.42	1.41	1.40	1.39	1.39
23	1.39	1.47	1.47	1.45	1.43	1.42	1.41	1.40	1.39	1.38
24	1.39	1.47	1.46	1.44	1.43	1.41	1.40	1.39	1.38	1.38
25	1.39	1.47	1.46	1.44	1.42	1.41	1.40	1.39	1.38	1.37
26	1.38	1.46	1.45	1.44	1.42	1.41	1.39	1.38	1.37	1.37
27	1.38	1.46	1.45	1.43	1.42	1.40	1.39	1.38	1.37	1.36
28	1.38	1.46	1.45	1.43	1.41	1.40	1.39	1.38	1.37	1.36
29	1.38	1.45	1.45	1.43	1.41	1.40	1.38	1.37	1.36	1.35
30	1.38	1.45	1.44	1.42	1.41	1.39	1.38	1.37	1.36	1.35
40	1.36	1.44	1.42	1.40	1.39	1.37	1.36	1.35	1.34	1.33
60	1.35	1.42	1.41	1.38	1.37	1.35	1.33	1.32	1.31	1.30
120	1.34	1.40	1.39	1.37	1.35	1.33	1.31	1.30	1.29	1.28
∞	1.32	1.39	1.37	1.35	1.33	1.31	1.29	1.28	1.27	1.25

Continued

	Percentages of the *F*-Distribution α = 0.25								
	Degrees of freedom for numerator ν1								
ν2	12	15	20	24	30	40	60	120	∞
1	9	9	10	10	10	10	10	10	10
2	3.39	3.41	3.43	3.43	3.44	3.45	3.46	3.47	3.48
3	2.45	2.46	2.46	2.46	2.47	2.47	2.47	2.47	2.47
4	2.08	2.08	2.08	2.08	2.08	2.08	2.08	2.08	2.08
5	1.89	1.89	1.88	1.88	1.88	1.88	1.87	1.87	1.87
6	1.77	1.76	1.76	1.75	1.75	1.75	1.74	1.74	1.74
7	1.68	1.68	1.67	1.67	1.66	1.66	1.65	1.65	1.65
8	1.62	1.62	1.61	1.60	1.60	1.59	1.59	1.58	1.58
9	1.58	1.57	1.56	1.56	1.55	1.54	1.54	1.53	1.53
10	1.54	1.53	1.52	1.52	1.51	1.51	1.50	1.49	1.48
11	1.51	1.50	1.49	1.49	1.48	1.47	1.47	1.46	1.45
12	1.49	1.48	1.47	1.46	1.45	1.45	1.44	1.43	1.42
13	1.47	1.46	1.45	1.44	1.43	1.42	1.42	1.41	1.40
14	1.45	1.44	1.43	1.42	1.41	1.41	1.40	1.39	1.38
15	1.44	1.43	1.41	1.41	1.40	1.39	1.38	1.37	1.36
16	1.43	1.41	1.40	1.39	1.38	1.37	1.36	1.35	1.34
17	1.41	1.40	1.39	1.38	1.37	1.36	1.35	1.34	1.33
18	1.40	1.39	1.38	1.37	1.36	1.35	1.34	1.33	1.32
19	1.40	1.38	1.37	1.36	1.35	1.34	1.33	1.32	1.30
20	1.39	1.37	1.36	1.35	1.34	1.33	1.32	1.31	1.29
21	1.38	1.37	1.35	1.34	1.33	1.32	1.31	1.30	1.28
22	1.37	1.36	1.34	1.33	1.32	1.31	1.30	1.29	1.28
23	1.37	1.35	1.34	1.33	1.32	1.31	1.30	1.28	1.27
24	1.36	1.35	1.33	1.32	1.31	1.30	1.29	1.28	1.26
25	1.36	1.34	1.33	1.32	1.31	1.29	1.28	1.27	1.25
26	1.35	1.34	1.32	1.31	1.30	1.29	1.28	1.26	1.25
27	1.35	1.33	1.32	1.31	1.30	1.28	1.27	1.26	1.24
28	1.34	1.33	1.31	1.30	1.29	1.28	1.27	1.25	1.24
29	1.34	1.32	1.31	1.30	1.29	1.27	1.26	1.25	1.23
30	1.34	1.32	1.30	1.29	1.28	1.27	1.26	1.24	1.23
40	1.31	1.30	1.28	1.26	1.25	1.24	1.22	1.21	1.19
60	1.29	1.27	1.25	1.24	1.22	1.21	1.19	1.17	1.15
120	1.26	1.24	1.22	1.21	1.19	1.18	1.16	1.13	1.10
∞	1.24	1.22	1.19	1.18	1.16	1.14	1.12	1.08	1.00

Appendix H
Tolerance Interval Factors

Tolerance interval factors—75%										
75% one-sided tolerance K_1 factors						75% two-sided tolerance K_2 factors				
75%	90%	95%	99%	99.9%	n	75%	90%	95%	99%	99.9%
2.225	3.992	5.122	7.267	9.672	2	4.498	6.301	7.414	9.531	11.920
1.464	2.501	3.152	4.396	5.805	3	2.501	3.538	4.187	5.431	6.844
1.255	2.134	2.681	3.726	4.911	4	2.035	2.892	3.431	4.471	5.657
1.152	1.962	2.463	3.421	4.507	5	1.825	2.599	3.088	4.033	5.117
1.088	1.859	2.336	3.244	4.273	6	1.704	2.429	2.889	3.779	4.802
1.043	1.790	2.250	3.126	4.119	7	1.624	2.318	2.757	3.611	4.593
1.010	1.740	2.188	3.042	4.008	8	1.568	2.238	2.663	3.491	4.444
0.985	1.701	2.141	2.977	3.924	9	1.525	2.178	2.593	3.400	4.330
0.964	1.671	2.104	2.927	3.858	10	1.492	2.131	2.537	3.328	4.241
0.947	1.645	2.073	2.885	3.804	11	1.465	2.093	2.493	3.271	4.169
0.932	1.624	2.048	2.851	3.760	12	1.443	2.062	2.456	3.223	4.110
0.920	1.606	2.026	2.822	3.722	13	1.425	2.036	2.424	3.183	4.059
0.909	1.591	2.007	2.797	3.690	14	1.409	2.013	2.398	3.148	4.016
0.899	1.577	1.991	2.775	3.661	15	1.395	1.994	2.375	3.118	3.979
0.891	1.565	1.976	2.756	3.636	16	1.383	1.977	2.355	3.092	3.946
0.883	1.554	1.963	2.739	3.614	17	1.372	1.962	2.337	3.069	3.917
0.876	1.545	1.952	2.723	3.595	18	1.363	1.948	2.321	3.048	3.891
0.870	1.536	1.941	2.710	3.577	19	1.355	1.936	2.307	3.030	3.867
0.864	1.528	1.932	2.697	3.560	20	1.347	1.925	2.294	3.013	3.846
0.859	1.521	1.923	2.685	3.546	21	1.340	1.915	2.282	2.998	3.827
0.854	1.514	1.915	2.675	3.532	22	1.334	1.906	2.271	2.984	3.809
0.849	1.508	1.908	2.665	3.520	23	1.328	1.898	2.261	2.971	3.793
0.845	1.502	1.901	2.656	3.508	24	1.322	1.891	2.252	2.959	3.778
0.841	1.497	1.895	2.648	3.497	25	1.317	1.883	2.244	2.948	3.764
0.825	1.475	1.869	2.614	3.453	30	1.297	1.855	2.210	2.904	3.708
0.813	1.458	1.849	2.588	3.421	35	1.283	1.834	2.185	2.871	3.667
0.803	1.445	1.834	2.568	3.395	40	1.271	1.818	2.166	2.846	3.635
0.788	1.425	1.811	2.538	3.357	50	1.255	1.794	2.138	2.809	3.588
0.777	1.411	1.795	2.517	3.331	60	1.243	1.778	2.118	2.784	3.556
0.763	1.392	1.772	2.489	3.294	80	1.228	1.756	2.092	2.749	3.512
0.753	1.380	1.758	2.470	3.270	100	1.218	1.742	2.075	2.727	3.484
0.709	1.324	1.693	2.364	3.143	500	1.177	1.683	2.006	2.636	3.368
0.698	1.295	1.663	2.353	3.127	1000	1.169	1.671	1.992	2.617	3.344
0.674	1.282	1.645	2.326	3.090	∞	1.150	1.645	1.960	2.576	3.291

Tolerance interval factors—90%										
90% one-sided tolerance K_1 factors						90% two-sided tolerance K_2 factors				
75%	90%	95%	99%	99.9%	n	75%	90%	95%	99%	99.9%
5.842	10.25	13.09	18.50	24.58	2	11.41	15.98	18.80	24.17	30.227
2.603	4.258	5.311	7.340	9.651	3	4.132	5.847	6.919	8.974	11.309
1.972	3.188	3.957	5.438	7.129	4	2.932	4.166	4.943	6.440	8.149
1.698	2.742	3.400	4.666	6.111	5	2.454	3.494	4.152	5.423	6.879
1.540	2.494	3.092	4.243	5.556	6	2.196	3.131	3.723	4.870	6.188
1.435	2.333	2.894	3.972	5.202	7	2.034	2.902	3.452	4.521	5.750
1.360	2.219	2.754	3.783	4.955	8	1.921	2.743	3.264	4.278	5.446
1.302	2.133	2.650	3.641	4.771	9	1.839	2.626	3.125	4.098	5.220
1.257	2.066	2.568	3.532	4.629	10	1.775	2.535	3.018	3.959	5.046
1.219	2.011	2.503	3.443	4.514	11	1.724	2.463	2.933	3.849	4.906
1.188	1.966	2.448	3.371	4.420	12	1.683	2.404	2.863	3.758	4.792
1.162	1.928	2.402	3.309	4.341	13	1.648	2.355	2.805	3.682	4.697
1.139	1.895	2.363	3.257	4.273	14	1.619	2.314	2.756	3.618	4.615
1.119	1.867	2.329	3.212	4.215	15	1.594	2.278	2.713	3.562	4.545
1.101	1.842	2.299	3.172	4.164	16	1.572	2.246	2.676	3.514	4.484
1.085	1.819	2.272	3.137	4.119	17	1.552	2.219	2.643	3.471	4.430
1.071	1.800	2.249	3.105	4.078	18	1.535	2.194	2.614	3.433	4.382
1.058	1.782	2.227	3.077	4.042	19	1.520	2.172	2.588	3.399	4.339
1.046	1.765	2.208	3.052	4.009	20	1.506	2.152	2.564	3.368	4.300
1.035	1.750	2.190	3.028	3.979	21	1.493	2.135	2.543	3.340	4.264
1.025	1.737	2.174	3.007	3.952	22	1.482	2.118	2.524	3.315	4.232
1.016	1.724	2.159	2.987	3.927	23	1.471	2.103	2.506	3.292	4.203
1.007	1.712	2.145	2.969	3.903	24	1.462	2.089	2.489	3.270	4.176
1.000	1.702	2.132	2.952	3.882	25	1.453	2.077	2.474	3.251	4.151
0.967	1.657	2.080	2.884	3.794	30	1.417	2.025	2.413	3.170	4.049
0.942	1.624	2.041	2.833	3.729	35	1.390	1.988	2.368	3.112	3.974
0.923	1.598	2.010	2.793	3.679	40	1.370	1.959	2.334	3.066	3.917
0.894	1.559	1.965	2.735	3.605	50	1.340	1.916	2.284	3.001	3.833
0.873	1.532	1.933	2.694	3.552	60	1.320	1.887	2.248	2.955	3.774
0.844	1.495	1.890	2.638	3.482	80	1.292	1.848	2.202	2.894	3.696
0.825	1.470	1.861	2.601	3.435	100	1.275	1.822	2.172	2.854	3.646
0.740	1.362	1.736	2.420	3.213	500	1.201	1.717	2.046	2.689	3.434
0.720	1.322	1.693	2.391	3.175	1000	1.185	1.695	2.019	2.654	3.390
0.674	1.282	1.645	2.326	3.090	∞	1.150	1.645	1.960	2.576	3.291

Tolerance interval factors—95%										
95% one-sided tolerance K_1 factors						95% two-sided tolerance K_2 factors				
75%	90%	95%	99%	99.9%	n	75%	90%	95%	99%	99.9%
11.76	20.58	26.26	37.09	49.28	2	22.86	32.02	37.67	48.43	60.573
3.806	6.155	7.656	10.55	13.86	3	5.922	8.380	9.916	12.86	16.208
2.618	4.162	5.144	7.042	9.214	4	3.779	5.369	6.370	8.299	10.502
2.150	3.407	4.203	5.741	7.502	5	3.002	4.275	5.079	6.634	8.415
1.895	3.006	3.708	5.062	6.612	6	2.604	3.712	4.414	5.775	7.337
1.732	2.755	3.399	4.642	6.063	7	2.361	3.369	4.007	5.248	6.676
1.618	2.582	3.187	4.354	5.688	8	2.197	3.136	3.732	4.891	6.226
1.532	2.454	3.031	4.143	5.413	9	2.078	2.967	3.532	4.631	5.899
1.465	2.355	2.911	3.981	5.203	10	1.987	2.839	3.379	4.433	5.649
1.411	2.275	2.815	3.852	5.036	11	1.916	2.737	3.259	4.277	5.452
1.366	2.210	2.736	3.747	4.900	12	1.858	2.655	3.162	4.150	5.291
1.328	2.155	2.671	3.659	4.787	13	1.810	2.587	3.081	4.044	5.158
1.296	2.109	2.614	3.585	4.690	14	1.770	2.529	3.012	3.955	5.045
1.268	2.068	2.566	3.520	4.607	15	1.735	2.480	2.954	3.878	4.949
1.243	2.033	2.524	3.464	4.535	16	1.705	2.437	2.903	3.812	4.865
1.220	2.002	2.486	3.414	4.471	17	1.679	2.400	2.858	3.754	4.791
1.201	1.974	2.453	3.370	4.415	18	1.655	2.366	2.819	3.702	4.725
1.183	1.949	2.423	3.331	4.364	19	1.635	2.337	2.784	3.656	4.667
1.166	1.926	2.396	3.295	4.318	20	1.616	2.310	2.752	3.615	4.614
1.152	1.905	2.371	3.263	4.277	21	1.599	2.286	2.723	3.577	4.567
1.138	1.886	2.349	3.233	4.239	22	1.584	2.264	2.697	3.543	4.523
1.125	1.869	2.328	3.206	4.204	23	1.570	2.244	2.673	3.512	4.484
1.114	1.853	2.309	3.181	4.172	24	1.557	2.225	2.651	3.483	4.447
1.103	1.838	2.292	3.158	4.142	25	1.545	2.208	2.631	3.457	4.413
1.058	1.777	2.220	3.064	4.022	30	1.497	2.140	2.549	3.350	4.278
1.025	1.732	2.167	2.995	3.934	35	1.462	2.090	2.490	3.272	4.179
0.999	1.697	2.125	2.941	3.865	40	1.435	2.052	2.445	3.212	4.103
0.960	1.646	2.065	2.862	3.766	50	1.396	1.996	2.379	3.126	3.993
0.933	1.609	2.022	2.807	3.695	60	1.369	1.958	2.333	3.066	3.916
0.895	1.559	1.964	2.733	3.601	80	1.334	1.907	2.272	2.986	3.814
0.870	1.527	1.927	2.684	3.539	100	1.311	1.874	2.233	2.934	3.748
0.758	1.385	1.763	2.454	3.255	500	1.215	1.737	2.070	2.721	3.475
0.733	1.338	1.712	2.415	3.205	1000	1.195	1.709	2.036	2.676	3.418
0.674	1.282	1.645	2.326	3.090	∞	1.150	1.645	1.960	2.576	3.291

Tolerance interval factors—99%										
99% one-sided tolerance K_1 factors						99% two-sided tolerance K_2 factors				
75%	90%	95%	99%	99.9%	*n*	75%	90%	95%	99%	99.9%
58.94	103.0	131.4	185.6	246.6	2	114.4	160.2	188.5	242.3	303.1
8.728	14.00	17.37	23.90	31.66	3	13.38	18.93	22.40	29.06	36.61
4.715	7.380	9.083	12.39	16.18	4	6.614	9.398	11.15	14.53	18.38
3.454	5.362	6.578	8.939	11.65	5	4.643	6.612	7.855	10.26	13.02
2.848	4.411	5.406	7.335	9.550	6	3.743	5.337	6.345	8.301	10.55
2.491	3.859	4.728	6.412	8.346	7	3.233	4.613	5.488	7.187	9.142
2.253	3.497	4.285	5.812	7.564	8	2.905	4.147	4.936	6.468	8.234
2.083	3.240	3.972	5.389	7.014	9	2.677	3.822	4.550	5.966	7.600
1.954	3.048	3.738	5.074	6.605	10	2.508	3.582	4.265	5.594	7.129
1.853	2.898	3.556	4.829	6.288	11	2.378	3.397	4.045	5.308	6.766
1.771	2.777	3.410	4.633	6.035	12	2.274	3.250	3.870	5.079	6.477
1.703	2.677	3.290	4.472	5.827	13	2.190	3.130	3.727	4.893	6.240
1.645	2.593	3.189	4.337	5.652	14	2.120	3.029	3.608	4.737	6.043
1.595	2.521	3.102	4.222	5.504	15	2.060	2.945	3.507	4.605	5.876
1.552	2.459	3.028	4.123	5.377	16	2.009	2.872	3.421	4.492	5.732
1.514	2.405	2.963	4.037	5.265	17	1.965	2.808	3.345	4.393	5.607
1.481	2.357	2.905	3.960	5.167	18	1.926	2.753	3.279	4.307	5.497
1.450	2.314	2.854	3.892	5.079	19	1.891	2.703	3.221	4.230	5.399
1.423	2.276	2.808	3.832	5.001	20	1.860	2.659	3.168	4.161	5.312
1.399	2.241	2.766	3.777	4.931	21	1.833	2.620	3.121	4.100	5.234
1.376	2.209	2.729	3.727	4.867	22	1.808	2.584	3.078	4.044	5.163
1.355	2.180	2.694	3.681	4.808	23	1.785	2.551	3.040	3.993	5.098
1.336	2.154	2.662	3.640	4.755	24	1.764	2.522	3.004	3.947	5.039
1.319	2.129	2.633	3.601	4.706	25	1.745	2.494	2.972	3.904	4.985
1.247	2.030	2.515	3.447	4.508	30	1.668	2.385	2.841	3.733	4.768
1.195	1.957	2.430	3.334	4.364	35	1.613	2.306	2.748	3.611	4.611
1.154	1.902	2.364	3.249	4.255	40	1.571	2.247	2.677	3.518	4.493
1.094	1.821	2.269	3.125	4.097	50	1.512	2.162	2.576	3.385	4.323
1.052	1.764	2.202	3.038	3.987	60	1.471	2.103	2.506	3.293	4.206
0.995	1.688	2.114	2.924	3.842	80	1.417	2.026	2.414	3.172	4.053
0.957	1.639	2.056	2.850	3.748	100	1.383	1.977	2.355	3.096	3.954
0.794	1.430	1.814	2.519	3.338	500	1.243	1.777	2.117	2.783	3.555
0.758	1.369	1.747	2.459	3.261	1000	1.214	1.736	2.068	2.718	3.472
0.674	1.282	1.645	2.326	3.090	∞	1.150	1.645	1.960	2.576	3.291

Appendix I

Critical Values of the Correlation Coeffecient

df	α values				df	α values			
(n − 2)	0.1	0.05	0.02	0.01	(n − 2)	0.1	0.05	0.02	0.01
1	.988	.997	1.000	1.000	21	.352	.413	.482	.526
2	.900	.950	.980	.990	22	.344	.404	.472	.515
3	.805	.878	.934	.959	23	.337	.396	.462	.505
4	.729	.811	.882	.917	24	.330	.388	.453	.496
5	.669	.754	.833	.874	25	.323	.381	.445	.487
6	.622	.707	.789	.834	26	.317	.374	.437	.479
7	.582	.666	.750	.798	27	.311	.367	.430	.471
8	.549	.632	.716	.765	28	.306	.361	.423	.463
9	.521	.602	.685	.735	29	.301	.355	.416	.456
10	.497	.576	.658	.708	30	.296	.349	.409	.449
11	.476	.553	.634	.684	35	.275	.325	.381	.418
12	.458	.532	.612	.661	40	.257	.304	.358	.393
13	.441	.514	.592	.641	45	.243	.288	.338	.372
14	.426	.497	.574	.623	50	.231	.273	.322	.354
15	.412	.482	.558	.606	60	.211	.250	.295	.325
16	.400	.468	.542	.590	70	.195	.232	.274	.303
17	.389	.456	.528	.575	80	.183	.217	.256	.283
18	.378	.444	.516	.561	90	.173	.205	.242	.267
19	.369	.433	.503	.549	100	.164	.195	.230	.254
20	.360	.423	.492	.537					

Appendix J

Critical Values of the Dean and Dixon Outlier Test

	$\alpha = 0.001$	$\alpha = 0.002$	$\alpha = 0.005$	$\alpha = 0.01$	$\alpha = 0.02$	$\alpha = 0.05$	$\alpha = 0.1$	$\alpha = 0.2$
n				r_{10}				
3	0.999	0.998	0.994	0.988	0.976	0.941	0.886	0.782
4	0.964	0.949	0.921	0.889	0.847	0.766	0.679	0.561
5	0.895	0.869	0.824	0.782	0.729	0.643	0.559	0.452
6	0.822	0.792	0.744	0.698	0.646	0.563	0.484	0.387
7	0.763	0.731	0.681	0.636	0.587	0.507	0.433	0.344
n				r_{11}				
8	0.799	0.769	0.724	0.682	0.633	0.554	0.48	0.386
9	0.75	0.72	0.675	0.634	0.586	0.512	0.441	0.352
10	0.713	0.683	0.637	0.597	0.551	0.477	0.409	0.325
n				r_{21}				
11	0.77	0.746	0.708	0.674	0.636	0.575	0.518	0.445
12	0.739	0.714	0.676	0.643	0.605	0.546	0.489	0.42
13	0.713	0.687	0.649	0.617	0.58	0.522	0.467	0.399
n				r_{22}				
14	0.732	0.708	0.672	0.64	0.603	0.546	0.491	0.422
15	0.708	0.685	0.648	0.617	0.582	0.524	0.47	0.403
16	0.691	0.667	0.63	0.598	0.562	0.505	0.453	0.386
17	0.671	0.647	0.611	0.58	0.545	0.489	0.437	0.373
18	0.652	0.628	0.594	0.564	0.529	0.475	0.424	0.361
19	0.64	0.617	0.581	0.551	0.517	0.462	0.412	0.349
20	0.627	0.604	0.568	0.538	0.503	0.45	0.401	0.339
25	0.574	0.55	0.517	0.489	0.457	0.406	0.359	0.302
30	0.539	0.517	0.484	0.456	0.425	0.376	0.332	0.278

Appendix K
Critical Values for the Grubbs Outlier Test

n	$\alpha = 0.001$	$\alpha = 0.005$	$\alpha = 0.01$	$\alpha = 0.025$	$\alpha = 0.05$	$\alpha = 0.1$
3	1.155	1.155	1.155	1.154	1.153	1.148
4	1.499	1.496	1.492	1.481	1.462	1.425
5	1.780	1.764	1.749	1.715	1.671	1.602
6	2.011	1.973	1.944	1.887	1.822	1.729
7	2.201	2.139	2.097	2.020	1.938	1.828
8	2.359	2.274	2.221	2.127	2.032	1.909
9	2.492	2.387	2.323	2.215	2.110	1.977
10	2.606	2.482	2.410	2.290	2.176	2.036
11	2.705	2.564	2.484	2.355	2.234	2.088
12	2.791	2.636	2.549	2.412	2.285	2.134
13	2.867	2.699	2.607	2.462	2.331	2.176
14	2.935	2.755	2.658	2.507	2.372	2.213
15	2.997	2.806	2.705	2.548	2.409	2.248
16	3.052	2.852	2.747	2.586	2.443	2.279
17	3.102	2.894	2.785	2.620	2.475	2.309
18	3.148	2.932	2.821	2.652	2.504	2.336
19	3.191	2.968	2.853	2.681	2.531	2.361
20	3.230	3.001	2.884	2.708	2.557	2.385
22	3.300	3.060	2.939	2.758	2.603	2.429
24	3.362	3.112	2.987	2.802	2.644	2.468
26	3.416	3.158	3.029	2.841	2.681	2.503
28	3.464	3.199	3.068	2.876	2.714	2.536
30	3.507	3.236	3.103	2.908	2.745	2.565
35	3.599	3.316	3.178	2.978	2.812	2.630
40	3.674	3.381	3.239	3.036	2.868	2.684
45	3.736	3.435	3.292	3.085	2.915	2.731
50	3.788	3.482	3.337	3.128	2.957	2.772

Continued

Continued

n	α = 0.001	α = 0.005	α = 0.01	α = 0.025	α = 0.05	α = 0.1
55	3.834	3.524	3.376	3.166	2.994	2.808
60	3.874	3.560	3.411	3.200	3.027	2.841
65	3.910	3.592	3.443	3.230	3.057	2.871
70	3.942	3.622	3.471	3.258	3.084	2.898
75	3.971	3.648	3.497	3.283	3.109	2.923
80	3.997	3.673	3.521	3.306	3.132	2.946
85	4.022	3.695	3.543	3.328	3.153	2.967
90	4.044	3.716	3.563	3.348	3.173	2.987
95	4.065	3.736	3.582	3.366	3.192	3.006
100	4.084	3.754	3.600	3.384	3.210	3.024
110	4.119	3.787	3.633	3.416	3.242	3.056
120	4.150	3.817	3.662	3.445	3.271	3.086
130	4.178	3.843	3.688	3.471	3.297	3.112
140	4.203	3.867	3.712	3.495	3.321	3.136
150	4.225	3.889	3.734	3.517	3.343	3.159
160	4.246	3.910	3.754	3.537	3.363	3.180
170	4.266	3.928	3.773	3.556	3.382	3.199
180	4.284	3.946	3.790	3.574	3.400	3.217
190	4.300	3.962	3.807	3.590	3.417	3.234
200	4.316	3.978	3.822	3.606	3.432	3.250
250	4.381	4.043	3.887	3.671	3.499	3.318
300	4.432	4.094	3.938	3.724	3.552	3.373
350	4.474	4.135	3.981	3.767	3.596	3.418
400	4.508	4.171	4.017	3.803	3.634	3.457
450	4.538	4.201	4.048	3.835	3.666	3.490
500	4.565	4.228	4.075	3.863	3.695	3.520
600	4.609	4.274	4.121	3.911	3.744	3.570
700	4.646	4.312	4.160	3.951	3.785	3.612
800	4.677	4.344	4.193	3.984	3.820	3.648
900	4.704	4.372	4.221	4.014	3.850	3.679
1000	4.728	4.397	4.247	4.040	3.877	3.707
1500	4.817	4.490	4.342	4.138	3.978	3.812
2000	4.878	4.553	4.407	4.206	4.048	3.884

Appendix L

Critical Values for the Discordance Outlier Test

n	α = 0.01	α = 0.05	n	α = 0.01	α = 0.05
3	1.155	1.153	27	3.049	2.698
4	1.492	1.463	28	3.068	2.714
5	1.749	1.672	29	3.085	2.730
6	1.944	1.822	30	3.103	2.745
7	2.097	1.938	31	3.119	2.759
8	2.221	2.032	32	3.135	2.773
9	2.323	2.110	33	3.150	2.786
10	2.410	2.176	34	3.164	2.799
11	2.485	2.234	35	3.178	2.811
12	2.550	2.285	36	3.191	2.823
13	2.607	2.331	37	3.204	2.835
14	2.659	2.371	38	3.216	2.846
15	2.705	2.409	39	3.228	2.857
16	2.747	2.443	40	3.240	2.866
17	2.785	2.475	41	3.251	2.877
18	2.821	2.504	42	3.261	2.887
19	2.854	2.532	43	3.271	2.896
20	2.884	2.557	44	3.282	2.905
21	2.912	2.580	45	3.292	2.914
22	2.939	2.603	46	3.302	2.923
23	2.963	2.624	47	3.310	2.931
24	2.987	2.644	48	3.319	2.940
25	3.009	2.663	49	3.329	2.948
26	3.029	2.681	50	3.336	2.956

Appendix M
The Binomial Distribution

n	r	0.01	0.015	0.02	0.025	0.03	0.035	0.04	0.045	0.05	0.10	0.15	0.20	0.25	0.30
2	0	0.9801	0.9702	0.9604	0.9506	0.9409	0.9312	0.9216	0.9120	0.9025	0.8100	0.7225	0.6400	0.5625	0.4900
	1	0.9999	0.9998	0.9996	0.9994	0.9991	0.9988	0.9984	0.9980	0.9975	0.9900	0.9775	0.9600	0.9375	0.9100
3	0	0.9703	0.9557	0.9412	0.9269	0.9127	0.8986	0.8847	0.8710	0.8574	0.7290	0.6141	0.5120	0.4219	0.3430
	1	0.9997	0.9993	0.9988	0.9982	0.9974	0.9964	0.9953	0.9941	0.9928	0.9720	0.9393	0.8960	0.8438	0.7840
	2	1.0000	1.0000	1.0000	1.0000	1.0000	1.0000	0.9999	0.9999	0.9999	0.9990	0.9966	0.9920	0.9844	0.9730
4	0	0.9606	0.9413	0.9224	0.9037	0.8853	0.8672	0.8493	0.8318	0.8145	0.6561	0.5220	0.4096	0.3164	0.2401
	1	0.9994	0.9987	0.9977	0.9964	0.9948	0.9930	0.9909	0.9886	0.9860	0.9477	0.8905	0.8192	0.7383	0.6517
	2	1.0000	1.0000	1.0000	0.9999	0.9999	0.9998	0.9998	0.9996	0.9995	0.9963	0.9880	0.9728	0.9492	0.9163
	3	1.0000	1.0000	1.0000	1.0000	1.0000	1.0000	1.0000	1.0000	1.0000	0.9999	0.9995	0.9984	0.9961	0.9919
5	0	0.9510	0.9272	0.9039	0.8811	0.8587	0.8368	0.8154	0.7944	0.7738	0.5905	0.4437	0.3277	0.2373	0.1681
	1	0.9990	0.9978	0.9962	0.9941	0.9915	0.9886	0.9852	0.9815	0.9774	0.9185	0.8352	0.7373	0.6328	0.5282
	2	1.0000	1.0000	0.9999	0.9998	0.9997	0.9996	0.9994	0.9991	0.9988	0.9914	0.9734	0.9421	0.8965	0.8369
	3	1.0000	1.0000	1.0000	1.0000	1.0000	1.0000	1.0000	1.0000	1.0000	0.9995	0.9978	0.9933	0.9844	0.9692
	4	1.0000	1.0000	1.0000	1.0000	1.0000	1.0000	1.0000	1.0000	1.0000	1.0000	0.9999	0.9997	0.9990	0.9976
6	0	0.9415	0.9133	0.8858	0.8591	0.8330	0.8075	0.7828	0.7586	0.7351	0.5314	0.3771	0.2621	0.1780	0.1176
	1	0.9985	0.9968	0.9943	0.9912	0.9875	0.9833	0.9784	0.9731	0.9672	0.8857	0.7765	0.6554	0.5339	0.4202
	2	1.0000	0.9999	0.9998	0.9997	0.9995	0.9992	0.9988	0.9984	0.9978	0.9842	0.9527	0.9011	0.8306	0.7443
	3	1.0000	1.0000	1.0000	1.0000	1.0000	1.0000	1.0000	0.9999	0.9999	0.9987	0.9941	0.9830	0.9624	0.9295
	4	1.0000	1.0000	1.0000	1.0000	1.0000	1.0000	1.0000	1.0000	1.0000	0.9999	0.9996	0.9984	0.9954	0.9891
	5	1.0000	1.0000	1.0000	1.0000	1.0000	1.0000	1.0000	1.0000	1.0000	1.0000	1.0000	0.9999	0.9998	0.9993
7	0	0.9321	0.8996	0.8681	0.8376	0.8080	0.7793	0.7514	0.7245	0.6983	0.4783	0.3206	0.2097	0.1335	0.0824
	1	0.9980	0.9955	0.9921	0.9879	0.9829	0.9771	0.9706	0.9634	0.9556	0.8503	0.7166	0.5767	0.4449	0.3294
	2	1.0000	0.9999	0.9997	0.9995	0.9991	0.9987	0.9980	0.9972	0.9962	0.9743	0.9262	0.8520	0.7564	0.6471
	3	1.0000	1.0000	1.0000	1.0000	1.0000	1.0000	0.9999	0.9999	0.9998	0.9973	0.9879	0.9667	0.9294	0.8740
	4	1.0000	1.0000	1.0000	1.0000	1.0000	1.0000	1.0000	1.0000	1.0000	0.9998	0.9988	0.9953	0.9871	0.9712
	5	1.0000	1.0000	1.0000	1.0000	1.0000	1.0000	1.0000	1.0000	1.0000	1.0000	0.9999	0.9996	0.9987	0.9962
	6	1.0000	1.0000	1.0000	1.0000	1.0000	1.0000	1.0000	1.0000	1.0000	1.0000	1.0000	1.0000	0.9999	0.9998
8	0	0.9227	0.8861	0.8508	0.8167	0.7837	0.7520	0.7214	0.6919	0.6634	0.4305	0.2725	0.1678	0.1001	0.0576
	1	0.9973	0.9941	0.9897	0.9842	0.9777	0.9702	0.9619	0.9527	0.9428	0.8131	0.6572	0.5033	0.3671	0.2553
	2	0.9999	0.9998	0.9996	0.9992	0.9987	0.9979	0.9969	0.9957	0.9942	0.9619	0.8948	0.7969	0.6785	0.5518
	3	1.0000	1.0000	1.0000	1.0000	0.9999	0.9999	0.9998	0.9998	0.9996	0.9950	0.9786	0.9437	0.8862	0.8059
	4	1.0000	1.0000	1.0000	1.0000	1.0000	1.0000	1.0000	1.0000	1.0000	0.9996	0.9971	0.9896	0.9727	0.9420
	5	1.0000	1.0000	1.0000	1.0000	1.0000	1.0000	1.0000	1.0000	1.0000	1.0000	0.9998	0.9988	0.9958	0.9887
	6	1.0000	1.0000	1.0000	1.0000	1.0000	1.0000	1.0000	1.0000	1.0000	1.0000	1.0000	0.9999	0.9996	0.9987
	7	1.0000	1.0000	1.0000	1.0000	1.0000	1.0000	1.0000	1.0000	1.0000	1.0000	1.0000	1.0000	1.0000	0.9999

Continued

n	r	0.35	0.40	0.45	0.50	0.55	0.60	0.65	0.70	0.75	0.80	0.85	0.90	0.95
2	0	0.4225	0.3600	0.3025	0.2500	0.2025	0.1600	0.1225	0.0900	0.0625	0.0400	0.0225	0.0100	0.0025
	1	0.8775	0.8400	0.7975	0.7500	0.6975	0.6400	0.5775	0.5100	0.4375	0.3600	0.2775	0.1900	0.0975
3	0	0.2746	0.2160	0.1664	0.1250	0.0911	0.0640	0.0429	0.0270	0.0156	0.0080	0.0034	0.0010	0.0001
	1	0.7183	0.6480	0.5748	0.5000	0.4253	0.3520	0.2818	0.2160	0.1563	0.1040	0.0608	0.0280	0.0073
	2	0.9571	0.9360	0.9089	0.8750	0.8336	0.7840	0.7254	0.6570	0.5781	0.4880	0.3859	0.2710	0.1426
4	0	0.1785	0.1296	0.0915	0.0625	0.0410	0.0256	0.0150	0.0081	0.0039	0.0016	0.0005	0.0001	0.0000
	1	0.5630	0.4752	0.3910	0.3125	0.2415	0.1792	0.1265	0.0837	0.0508	0.0272	0.0120	0.0037	0.0005
	2	0.8735	0.8208	0.7585	0.6875	0.6090	0.5248	0.4370	0.3483	0.2617	0.1808	0.1095	0.0523	0.0140
	3	0.9850	0.9744	0.9590	0.9375	0.9085	0.8704	0.8215	0.7599	0.6836	0.5904	0.4780	0.3439	0.1855
5	0	0.1160	0.0778	0.0503	0.0313	0.0185	0.0102	0.0053	0.0024	0.0010	0.0003	0.0001	0.0000	0.0000
	1	0.4284	0.3370	0.2562	0.1875	0.1312	0.0870	0.0540	0.0308	0.0156	0.0067	0.0022	0.0005	0.0000
	2	0.7648	0.6826	0.5931	0.5000	0.4069	0.3174	0.2352	0.1631	0.1035	0.0579	0.0266	0.0086	0.0012
	3	0.9460	0.9130	0.8688	0.8125	0.7438	0.6630	0.5716	0.4718	0.3672	0.2627	0.1648	0.0815	0.0226
	4	0.9947	0.9898	0.9815	0.9688	0.9497	0.9222	0.8840	0.8319	0.7627	0.6723	0.5563	0.4095	0.2262
6	0	0.0754	0.0467	0.0277	0.0156	0.0083	0.0041	0.0018	0.0007	0.0002	0.0001	0.0000	0.0000	0.0000
	1	0.3191	0.2333	0.1636	0.1094	0.0692	0.0410	0.0223	0.0109	0.0046	0.0016	0.0004	0.0001	0.0000
	2	0.6471	0.5443	0.4415	0.3438	0.2553	0.1792	0.1174	0.0705	0.0376	0.0170	0.0059	0.0013	0.0001
	3	0.8826	0.8208	0.7447	0.6563	0.5585	0.4557	0.3529	0.2557	0.1694	0.0989	0.0473	0.0159	0.0022
	4	0.9777	0.9590	0.9308	0.8906	0.8364	0.7667	0.6809	0.5798	0.4661	0.3446	0.2235	0.1143	0.0328
	5	0.9982	0.9959	0.9917	0.9844	0.9723	0.9533	0.9246	0.8824	0.8220	0.7379	0.6229	0.4686	0.2649
7	0	0.0490	0.0280	0.0152	0.0078	0.0037	0.0016	0.0006	0.0002	0.0001	0.0000	0.0000	0.0000	0.0000
	1	0.2338	0.1586	0.1024	0.0625	0.0357	0.0188	0.0090	0.0038	0.0013	0.0004	0.0001	0.0000	0.0000
	2	0.5323	0.4199	0.3164	0.2266	0.1529	0.0963	0.0556	0.0288	0.0129	0.0047	0.0012	0.0002	0.0000
	3	0.8002	0.7102	0.6083	0.5000	0.3917	0.2898	0.1998	0.1260	0.0706	0.0333	0.0121	0.0027	0.0002
	4	0.9444	0.9037	0.8471	0.7734	0.6836	0.5801	0.4677	0.3529	0.2436	0.1480	0.0738	0.0257	0.0038
	5	0.9910	0.9812	0.9643	0.9375	0.8976	0.8414	0.7662	0.6706	0.5551	0.4233	0.2834	0.1497	0.0444
	6	0.9994	0.9984	0.9963	0.9922	0.9848	0.9720	0.9510	0.9176	0.8665	0.7903	0.6794	0.5217	0.3017
8	0	0.0319	0.0168	0.0084	0.0039	0.0017	0.0007	0.0002	0.0001	0.0000	0.0000	0.0000	0.0000	0.0000
	1	0.1691	0.1064	0.0632	0.0352	0.0181	0.0085	0.0036	0.0013	0.0004	0.0001	0.0000	0.0000	0.0000
	2	0.4278	0.3154	0.2201	0.1445	0.0885	0.0498	0.0253	0.0113	0.0042	0.0012	0.0002	0.0000	0.0000
	3	0.7064	0.5941	0.4770	0.3633	0.2604	0.1737	0.1061	0.0580	0.0273	0.0104	0.0029	0.0004	0.0000
	4	0.8939	0.8263	0.7396	0.6367	0.5230	0.4059	0.2936	0.1941	0.1138	0.0563	0.0214	0.0050	0.0004
	5	0.9747	0.9502	0.9115	0.8555	0.7799	0.6846	0.5722	0.4482	0.3215	0.2031	0.1052	0.0381	0.0058
	6	0.9964	0.9915	0.9819	0.9648	0.9368	0.8936	0.8309	0.7447	0.6329	0.4967	0.3428	0.1869	0.0572
	7	0.9998	0.9993	0.9983	0.9961	0.9916	0.9832	0.9681	0.9424	0.8999	0.8322	0.7275	0.5695	0.3366

Continued

n	r	0.01	0.015	0.02	0.025	0.03	0.035	0.04	0.045	0.05	0.10	0.15	0.20	0.25	0.30
9	0	0.9135	0.8728	0.8337	0.7962	0.7602	0.7257	0.6925	0.6607	0.6302	0.3874	0.2316	0.1342	0.0751	0.0404
	1	0.9966	0.9924	0.9869	0.9800	0.9718	0.9626	0.9522	0.9409	0.9288	0.7748	0.5995	0.4362	0.3003	0.1960
	2	0.9999	0.9997	0.9994	0.9988	0.9980	0.9969	0.9955	0.9938	0.9916	0.9470	0.8591	0.7382	0.6007	0.4628
	3	1.0000	1.0000	1.0000	1.0000	0.9999	0.9998	0.9997	0.9996	0.9994	0.9917	0.9661	0.9144	0.8343	0.7297
	4	1.0000	1.0000	1.0000	1.0000	1.0000	1.0000	1.0000	1.0000	1.0000	0.9991	0.9944	0.9804	0.9511	0.9012
	5	1.0000	1.0000	1.0000	1.0000	1.0000	1.0000	1.0000	1.0000	1.0000	0.9999	0.9994	0.9969	0.9900	0.9747
	6	1.0000	1.0000	1.0000	1.0000	1.0000	1.0000	1.0000	1.0000	1.0000	1.0000	1.0000	0.9997	0.9987	0.9957
	7	1.0000	1.0000	1.0000	1.0000	1.0000	1.0000	1.0000	1.0000	1.0000	1.0000	1.0000	1.0000	0.9999	0.9996
	8	1.0000	1.0000	1.0000	1.0000	1.0000	1.0000	1.0000	1.0000	1.0000	1.0000	1.0000	1.0000	1.0000	1.0000
10	0	0.9044	0.8597	0.8171	0.7763	0.7374	0.7003	0.6648	0.6310	0.5987	0.3487	0.1969	0.1074	0.0563	0.0282
	1	0.9957	0.9907	0.9838	0.9754	0.9655	0.9543	0.9418	0.9283	0.9139	0.7361	0.5443	0.3758	0.2440	0.1493
	2	0.9999	0.9996	0.9991	0.9984	0.9972	0.9957	0.9938	0.9914	0.9885	0.9298	0.8202	0.6778	0.5256	0.3828
	3	1.0000	1.0000	1.0000	0.9999	0.9999	0.9997	0.9996	0.9993	0.9990	0.9872	0.9500	0.8791	0.7759	0.6496
	4	1.0000	1.0000	1.0000	1.0000	1.0000	1.0000	1.0000	1.0000	0.9999	0.9984	0.9901	0.9672	0.9219	0.8497
	5	1.0000	1.0000	1.0000	1.0000	1.0000	1.0000	1.0000	1.0000	1.0000	0.9999	0.9986	0.9936	0.9803	0.9527
	6	1.0000	1.0000	1.0000	1.0000	1.0000	1.0000	1.0000	1.0000	1.0000	1.0000	0.9999	0.9991	0.9965	0.9894
	7	1.0000	1.0000	1.0000	1.0000	1.0000	1.0000	1.0000	1.0000	1.0000	1.0000	1.0000	0.9999	0.9996	0.9984
	8	1.0000	1.0000	1.0000	1.0000	1.0000	1.0000	1.0000	1.0000	1.0000	1.0000	1.0000	1.0000	1.0000	0.9999
	9	1.0000	1.0000	1.0000	1.0000	1.0000	1.0000	1.0000	1.0000	1.0000	1.0000	1.0000	1.0000	1.0000	1.0000
11	0	0.8953	0.8468	0.8007	0.7569	0.7153	0.6758	0.6382	0.6026	0.5688	0.3138	0.1673	0.0859	0.0422	0.0198
	1	0.9948	0.9887	0.9805	0.9704	0.9587	0.9454	0.9308	0.9150	0.8981	0.6974	0.4922	0.3221	0.1971	0.1130
	2	0.9998	0.9995	0.9988	0.9978	0.9963	0.9943	0.9917	0.9885	0.9848	0.9104	0.7788	0.6174	0.4552	0.3127
	3	1.0000	1.0000	1.0000	0.9999	0.9998	0.9996	0.9993	0.9990	0.9984	0.9815	0.9306	0.8389	0.7133	0.5696
	4	1.0000	1.0000	1.0000	1.0000	1.0000	1.0000	1.0000	0.9999	0.9999	0.9972	0.9841	0.9496	0.8854	0.7897
	5	1.0000	1.0000	1.0000	1.0000	1.0000	1.0000	1.0000	1.0000	1.0000	0.9997	0.9973	0.9883	0.9657	0.9218
	6	1.0000	1.0000	1.0000	1.0000	1.0000	1.0000	1.0000	1.0000	1.0000	1.0000	0.9997	0.9980	0.9924	0.9784
	7	1.0000	1.0000	1.0000	1.0000	1.0000	1.0000	1.0000	1.0000	1.0000	1.0000	1.0000	0.9998	0.9988	0.9957
	8	1.0000	1.0000	1.0000	1.0000	1.0000	1.0000	1.0000	1.0000	1.0000	1.0000	1.0000	1.0000	0.9999	0.9994
	9	1.0000	1.0000	1.0000	1.0000	1.0000	1.0000	1.0000	1.0000	1.0000	1.0000	1.0000	1.0000	1.0000	1.0000
	10	1.0000	1.0000	1.0000	1.0000	1.0000	1.0000	1.0000	1.0000	1.0000	1.0000	1.0000	1.0000	1.0000	1.0000

Continued

n	r	0.35	0.40	0.45	0.50	0.55	0.60	0.65	0.70	0.75	0.80	0.85	0.90	0.95
9	0	0.0207	0.0101	0.0046	0.0020	0.0008	0.0003	0.0001	0.0000	0.0000	0.0000	0.0000	0.0000	0.0000
	1	0.1211	0.0705	0.0385	0.0195	0.0091	0.0038	0.0014	0.0004	0.0001	0.0000	0.0000	0.0000	0.0000
	2	0.3373	0.2318	0.1495	0.0898	0.0498	0.0250	0.0112	0.0043	0.0013	0.0003	0.0000	0.0000	0.0000
	3	0.6089	0.4826	0.3614	0.2539	0.1658	0.0994	0.0536	0.0253	0.0100	0.0031	0.0006	0.0001	0.0000
	4	0.8283	0.7334	0.6214	0.5000	0.3786	0.2666	0.1717	0.0988	0.0489	0.0196	0.0056	0.0009	0.0000
	5	0.9464	0.9006	0.8342	0.7461	0.6386	0.5174	0.3911	0.2703	0.1657	0.0856	0.0339	0.0083	0.0006
	6	0.9888	0.9750	0.9502	0.9102	0.8505	0.7682	0.6627	0.5372	0.3993	0.2618	0.1409	0.0530	0.0084
	7	0.9986	0.9962	0.9909	0.9805	0.9615	0.9295	0.8789	0.8040	0.6997	0.5638	0.4005	0.2252	0.0712
	8	0.9999	0.9997	0.9992	0.9980	0.9954	0.9899	0.9793	0.9596	0.9249	0.8658	0.7684	0.6126	0.3698
10	0	0.0135	0.0060	0.0025	0.0010	0.0003	0.0001	0.0000	0.0000	0.0000	0.0000	0.0000	0.0000	0.0000
	1	0.0860	0.0464	0.0233	0.0107	0.0045	0.0017	0.0005	0.0001	0.0000	0.0000	0.0000	0.0000	0.0000
	2	0.2616	0.1673	0.0996	0.0547	0.0274	0.0123	0.0048	0.0016	0.0004	0.0001	0.0000	0.0000	0.0000
	3	0.5138	0.3823	0.2660	0.1719	0.1020	0.0548	0.0260	0.0106	0.0035	0.0009	0.0001	0.0000	0.0000
	4	0.7515	0.6331	0.5044	0.3770	0.2616	0.1662	0.0949	0.0473	0.0197	0.0064	0.0014	0.0001	0.0000
	5	0.9051	0.8338	0.7384	0.6230	0.4956	0.3669	0.2485	0.1503	0.0781	0.0328	0.0099	0.0016	0.0001
	6	0.9740	0.9452	0.8980	0.8281	0.7340	0.6177	0.4862	0.3504	0.2241	0.1209	0.0500	0.0128	0.0010
	7	0.9952	0.9877	0.9726	0.9453	0.9004	0.8327	0.7384	0.6172	0.4744	0.3222	0.1798	0.0702	0.0115
	8	0.9995	0.9983	0.9955	0.9893	0.9767	0.9536	0.9140	0.8507	0.7560	0.6242	0.4557	0.2639	0.0861
	9	1.0000	0.9999	0.9997	0.9990	0.9975	0.9940	0.9865	0.9718	0.9437	0.8926	0.8031	0.6513	0.4013
11	0	0.0088	0.0036	0.0014	0.0005	0.0002	0.0000	0.0000	0.0000	0.0000	0.0000	0.0000	0.0000	0.0000
	1	0.0606	0.0302	0.0139	0.0059	0.0022	0.0007	0.0002	0.0000	0.0000	0.0000	0.0000	0.0000	0.0000
	2	0.2001	0.1189	0.0652	0.0327	0.0148	0.0059	0.0020	0.0006	0.0001	0.0000	0.0000	0.0000	0.0000
	3	0.4256	0.2963	0.1911	0.1133	0.0610	0.0293	0.0122	0.0043	0.0012	0.0002	0.0000	0.0000	0.0000
	4	0.6683	0.5328	0.3971	0.2744	0.1738	0.0994	0.0501	0.0216	0.0076	0.0020	0.0003	0.0000	0.0000
	5	0.8513	0.7535	0.6331	0.5000	0.3669	0.2465	0.1487	0.0782	0.0343	0.0117	0.0027	0.0003	0.0000
	6	0.9499	0.9006	0.8262	0.7256	0.6029	0.4672	0.3317	0.2103	0.1146	0.0504	0.0159	0.0028	0.0001
	7	0.9878	0.9707	0.9390	0.8867	0.8089	0.7037	0.5744	0.4304	0.2867	0.1611	0.0694	0.0185	0.0016
	8	0.9980	0.9941	0.9852	0.9673	0.9348	0.8811	0.7999	0.6873	0.5448	0.3826	0.2212	0.0896	0.0152
	9	0.9998	0.9993	0.9978	0.9941	0.9861	0.9698	0.9394	0.8870	0.8029	0.6779	0.5078	0.3026	0.1019
	10	1.0000	1.0000	0.9998	0.9995	0.9986	0.9964	0.9912	0.9802	0.9578	0.9141	0.8327	0.6862	0.4312

Continued

n	r	0.01	0.015	0.02	0.025	0.03	0.035	0.04	0.045	0.05	0.10	0.15	0.20	0.25	0.30
12	0	0.8864	0.8341	0.7847	0.7380	0.6938	0.6521	0.6127	0.5755	0.5404	0.2824	0.1422	0.0687	0.0317	0.0138
	1	0.9938	0.9866	0.9769	0.9651	0.9514	0.9359	0.9191	0.9009	0.8816	0.6590	0.4435	0.2749	0.1584	0.0850
	2	0.9998	0.9993	0.9985	0.9971	0.9952	0.9926	0.9893	0.9852	0.9804	0.8891	0.7358	0.5583	0.3907	0.2528
	3	1.0000	1.0000	0.9999	0.9998	0.9997	0.9994	0.9990	0.9985	0.9978	0.9744	0.9078	0.7946	0.6488	0.4925
	4	1.0000	1.0000	1.0000	1.0000	1.0000	1.0000	0.9999	0.9999	0.9998	0.9957	0.9761	0.9274	0.8424	0.7237
	5	1.0000	1.0000	1.0000	1.0000	1.0000	1.0000	1.0000	1.0000	1.0000	0.9995	0.9954	0.9806	0.9456	0.8822
	6	1.0000	1.0000	1.0000	1.0000	1.0000	1.0000	1.0000	1.0000	1.0000	0.9999	0.9993	0.9961	0.9857	0.9614
	7	1.0000	1.0000	1.0000	1.0000	1.0000	1.0000	1.0000	1.0000	1.0000	1.0000	0.9999	0.9994	0.9972	0.9905
	8	1.0000	1.0000	1.0000	1.0000	1.0000	1.0000	1.0000	1.0000	1.0000	1.0000	1.0000	0.9999	0.9996	0.9983
	9	1.0000	1.0000	1.0000	1.0000	1.0000	1.0000	1.0000	1.0000	1.0000	1.0000	1.0000	1.0000	1.0000	0.9998
	10	1.0000	1.0000	1.0000	1.0000	1.0000	1.0000	1.0000	1.0000	1.0000	1.0000	1.0000	1.0000	1.0000	1.0000
	11	1.0000	1.0000	1.0000	1.0000	1.0000	1.0000	1.0000	1.0000	1.0000	1.0000	1.0000	1.0000	1.0000	1.0000
13	0	0.8775	0.8216	0.7690	0.7195	0.6730	0.6293	0.5882	0.5496	0.5133	0.2542	0.1209	0.0550	0.0238	0.0097
	1	0.9928	0.9843	0.9730	0.9594	0.9436	0.9260	0.9068	0.8863	0.8646	0.6213	0.3983	0.2336	0.1267	0.0637
	2	0.9997	0.9991	0.9980	0.9963	0.9938	0.9906	0.9865	0.9814	0.9755	0.8661	0.6920	0.5017	0.3326	0.2025
	3	1.0000	1.0000	0.9999	0.9998	0.9995	0.9992	0.9986	0.9979	0.9969	0.9658	0.8820	0.7473	0.5843	0.4206
	4	1.0000	1.0000	1.0000	1.0000	1.0000	0.9999	0.9999	0.9998	0.9997	0.9935	0.9658	0.9009	0.7940	0.6543
	5	1.0000	1.0000	1.0000	1.0000	1.0000	1.0000	1.0000	1.0000	1.0000	0.9991	0.9925	0.9700	0.9198	0.8346
	6	1.0000	1.0000	1.0000	1.0000	1.0000	1.0000	1.0000	1.0000	1.0000	0.9999	0.9987	0.9930	0.9757	0.9376
	7	1.0000	1.0000	1.0000	1.0000	1.0000	1.0000	1.0000	1.0000	1.0000	1.0000	0.9998	0.9988	0.9944	0.9818
	8	1.0000	1.0000	1.0000	1.0000	1.0000	1.0000	1.0000	1.0000	1.0000	1.0000	1.0000	0.9998	0.9990	0.9960
	9	1.0000	1.0000	1.0000	1.0000	1.0000	1.0000	1.0000	1.0000	1.0000	1.0000	1.0000	1.0000	0.9999	0.9993
	10	1.0000	1.0000	1.0000	1.0000	1.0000	1.0000	1.0000	1.0000	1.0000	1.0000	1.0000	1.0000	1.0000	0.9999
	11	1.0000	1.0000	1.0000	1.0000	1.0000	1.0000	1.0000	1.0000	1.0000	1.0000	1.0000	1.0000	1.0000	1.0000
	12	1.0000	1.0000	1.0000	1.0000	1.0000	1.0000	1.0000	1.0000	1.0000	1.0000	1.0000	1.0000	1.0000	1.0000

Continued

n	r	0.35	0.40	0.45	0.50	0.55	0.60	0.65	0.70	0.75	0.80	0.85	0.90	0.95
12	0	0.0057	0.0022	0.0008	0.0002	0.0001	0.0000	0.0000	0.0000	0.0000	0.0000	0.0000	0.0000	0.0000
	1	0.0424	0.0196	0.0083	0.0032	0.0011	0.0003	0.0001	0.0000	0.0000	0.0000	0.0000	0.0000	0.0000
	2	0.1513	0.0834	0.0421	0.0193	0.0079	0.0028	0.0008	0.0002	0.0000	0.0000	0.0000	0.0000	0.0000
	3	0.3467	0.2253	0.1345	0.0730	0.0356	0.0153	0.0056	0.0017	0.0004	0.0001	0.0000	0.0000	0.0000
	4	0.5833	0.4382	0.3044	0.1938	0.1117	0.0573	0.0255	0.0095	0.0028	0.0006	0.0001	0.0000	0.0000
	5	0.7873	0.6652	0.5269	0.3872	0.2607	0.1582	0.0846	0.0386	0.0143	0.0039	0.0007	0.0001	0.0000
	6	0.9154	0.8418	0.7393	0.6128	0.4731	0.3348	0.2127	0.1178	0.0544	0.0194	0.0046	0.0005	0.0000
	7	0.9745	0.9427	0.8883	0.8062	0.6956	0.5618	0.4167	0.2763	0.1576	0.0726	0.0239	0.0043	0.0002
	8	0.9944	0.9847	0.9644	0.9270	0.8655	0.7747	0.6533	0.5075	0.3512	0.2054	0.0922	0.0256	0.0022
	9	0.9992	0.9972	0.9921	0.9807	0.9579	0.9166	0.8487	0.7472	0.6093	0.4417	0.2642	0.1109	0.0196
	10	0.9999	0.9997	0.9989	0.9968	0.9917	0.9804	0.9576	0.9150	0.8416	0.7251	0.5565	0.3410	0.1184
	11	1.0000	1.0000	0.9999	0.9998	0.9992	0.9978	0.9943	0.9862	0.9683	0.9313	0.8578	0.7176	0.4596
13	0	0.0037	0.0013	0.0004	0.0001	0.0000	0.0000	0.0000	0.0000	0.0000	0.0000	0.0000	0.0000	0.0000
	1	0.0296	0.0126	0.0049	0.0017	0.0005	0.0001	0.0000	0.0000	0.0000	0.0000	0.0000	0.0000	0.0000
	2	0.1132	0.0579	0.0269	0.0112	0.0041	0.0013	0.0003	0.0001	0.0000	0.0000	0.0000	0.0000	0.0000
	3	0.2783	0.1686	0.0929	0.0461	0.0203	0.0078	0.0025	0.0007	0.0001	0.0000	0.0000	0.0000	0.0000
	4	0.5005	0.3530	0.2279	0.1334	0.0698	0.0321	0.0126	0.0040	0.0010	0.0002	0.0000	0.0000	0.0000
	5	0.7159	0.5744	0.4268	0.2905	0.1788	0.0977	0.0462	0.0182	0.0056	0.0012	0.0002	0.0000	0.0000
	6	0.8705	0.7712	0.6437	0.5000	0.3563	0.2288	0.1295	0.0624	0.0243	0.0070	0.0013	0.0001	0.0000
	7	0.9538	0.9023	0.8212	0.7095	0.5732	0.4256	0.2841	0.1654	0.0802	0.0300	0.0075	0.0009	0.0000
	8	0.9874	0.9679	0.9302	0.8666	0.7721	0.6470	0.4995	0.3457	0.2060	0.0991	0.0342	0.0065	0.0003
	9	0.9975	0.9922	0.9797	0.9539	0.9071	0.8314	0.7217	0.5794	0.4157	0.2527	0.1180	0.0342	0.0031
	10	0.9997	0.9987	0.9959	0.9888	0.9731	0.9421	0.8868	0.7975	0.6674	0.4983	0.3080	0.1339	0.0245
	11	1.0000	0.9999	0.9995	0.9983	0.9951	0.9874	0.9704	0.9363	0.8733	0.7664	0.6017	0.3787	0.1354
	12	1.0000	1.0000	1.0000	0.9999	0.9996	0.9987	0.9963	0.9903	0.9762	0.9450	0.8791	0.7458	0.4867

Continued

n	r	0.01	0.015	0.02	0.025	0.03	0.035	0.04	0.045	0.05	0.10	0.15	0.20	0.25	0.30
14	0	0.8687	0.8093	0.7536	0.7016	0.6528	0.6073	0.5647	0.5249	0.4877	0.2288	0.1028	0.0440	0.0178	0.0068
	1	0.9916	0.9818	0.9690	0.9534	0.9355	0.9156	0.8941	0.8711	0.8470	0.5846	0.3567	0.1979	0.1010	0.0475
	2	0.9997	0.9989	0.9975	0.9954	0.9923	0.9883	0.9833	0.9772	0.9699	0.8416	0.6479	0.4481	0.2811	0.1608
	3	1.0000	1.0000	0.9999	0.9997	0.9994	0.9989	0.9981	0.9971	0.9958	0.9559	0.8535	0.6982	0.5213	0.3552
	4	1.0000	1.0000	1.0000	1.0000	1.0000	0.9999	0.9998	0.9997	0.9996	0.9908	0.9533	0.8702	0.7415	0.5842
	5	1.0000	1.0000	1.0000	1.0000	1.0000	1.0000	1.0000	1.0000	1.0000	0.9985	0.9885	0.9561	0.8883	0.7805
	6	1.0000	1.0000	1.0000	1.0000	1.0000	1.0000	1.0000	1.0000	1.0000	0.9998	0.9978	0.9884	0.9617	0.9067
	7	1.0000	1.0000	1.0000	1.0000	1.0000	1.0000	1.0000	1.0000	1.0000	1.0000	0.9997	0.9976	0.9897	0.9685
	8	1.0000	1.0000	1.0000	1.0000	1.0000	1.0000	1.0000	1.0000	1.0000	1.0000	1.0000	0.9996	0.9978	0.9917
	9	1.0000	1.0000	1.0000	1.0000	1.0000	1.0000	1.0000	1.0000	1.0000	1.0000	1.0000	1.0000	0.9997	0.9983
	10	1.0000	1.0000	1.0000	1.0000	1.0000	1.0000	1.0000	1.0000	1.0000	1.0000	1.0000	1.0000	1.0000	0.9998
	11	1.0000	1.0000	1.0000	1.0000	1.0000	1.0000	1.0000	1.0000	1.0000	1.0000	1.0000	1.0000	1.0000	1.0000
	12	1.0000	1.0000	1.0000	1.0000	1.0000	1.0000	1.0000	1.0000	1.0000	1.0000	1.0000	1.0000	1.0000	1.0000
	13	1.0000	1.0000	1.0000	1.0000	1.0000	1.0000	1.0000	1.0000	1.0000	1.0000	1.0000	1.0000	1.0000	1.0000
15	0	0.8601	0.7972	0.7386	0.6840	0.6333	0.5860	0.5421	0.5012	0.4633	0.2059	0.0874	0.0352	0.0134	0.0047
	1	0.9904	0.9792	0.9647	0.9471	0.9270	0.9048	0.8809	0.8555	0.8290	0.5490	0.3186	0.1671	0.0802	0.0353
	2	0.9996	0.9987	0.9970	0.9943	0.9906	0.9858	0.9797	0.9724	0.9638	0.8159	0.6042	0.3980	0.2361	0.1268
	3	1.0000	0.9999	0.9998	0.9996	0.9992	0.9985	0.9976	0.9962	0.9945	0.9444	0.8227	0.6482	0.4613	0.2969
	4	1.0000	1.0000	1.0000	1.0000	0.9999	0.9999	0.9998	0.9996	0.9994	0.9873	0.9383	0.8358	0.6865	0.5155
	5	1.0000	1.0000	1.0000	1.0000	1.0000	1.0000	1.0000	1.0000	0.9999	0.9978	0.9832	0.9389	0.8516	0.7216
	6	1.0000	1.0000	1.0000	1.0000	1.0000	1.0000	1.0000	1.0000	1.0000	0.9997	0.9964	0.9819	0.9434	0.8689
	7	1.0000	1.0000	1.0000	1.0000	1.0000	1.0000	1.0000	1.0000	1.0000	1.0000	0.9994	0.9958	0.9827	0.9500
	8	1.0000	1.0000	1.0000	1.0000	1.0000	1.0000	1.0000	1.0000	1.0000	1.0000	0.9999	0.9992	0.9958	0.9848
	9	1.0000	1.0000	1.0000	1.0000	1.0000	1.0000	1.0000	1.0000	1.0000	1.0000	1.0000	0.9999	0.9992	0.9963
	10	1.0000	1.0000	1.0000	1.0000	1.0000	1.0000	1.0000	1.0000	1.0000	1.0000	1.0000	1.0000	0.9999	0.9993
	11	1.0000	1.0000	1.0000	1.0000	1.0000	1.0000	1.0000	1.0000	1.0000	1.0000	1.0000	1.0000	1.0000	0.9999
	12	1.0000	1.0000	1.0000	1.0000	1.0000	1.0000	1.0000	1.0000	1.0000	1.0000	1.0000	1.0000	1.0000	1.0000
	13	1.0000	1.0000	1.0000	1.0000	1.0000	1.0000	1.0000	1.0000	1.0000	1.0000	1.0000	1.0000	1.0000	1.0000
	14	1.0000	1.0000	1.0000	1.0000	1.0000	1.0000	1.0000	1.0000	1.0000	1.0000	1.0000	1.0000	1.0000	1.0000

Continued

n	r	0.35	0.40	0.45	0.50	0.55	0.60	0.65	0.70	0.75	0.80	0.85	0.90	0.95
14	0	0.0024	0.0008	0.0002	0.0001	0.0000	0.0000	0.0000	0.0000	0.0000	0.0000	0.0000	0.0000	0.0000
	1	0.0205	0.0081	0.0029	0.0009	0.0003	0.0001	0.0000	0.0000	0.0000	0.0000	0.0000	0.0000	0.0000
	2	0.0839	0.0398	0.0170	0.0065	0.0022	0.0006	0.0001	0.0000	0.0000	0.0000	0.0000	0.0000	0.0000
	3	0.2205	0.1243	0.0632	0.0287	0.0114	0.0039	0.0011	0.0002	0.0000	0.0000	0.0000	0.0000	0.0000
	4	0.4227	0.2793	0.1672	0.0898	0.0426	0.0175	0.0060	0.0017	0.0003	0.0000	0.0000	0.0000	0.0000
	5	0.6405	0.4859	0.3373	0.2120	0.1189	0.0583	0.0243	0.0083	0.0022	0.0004	0.0000	0.0000	0.0000
	6	0.8164	0.6925	0.5461	0.3953	0.2586	0.1501	0.0753	0.0315	0.0103	0.0024	0.0003	0.0000	0.0000
	7	0.9247	0.8499	0.7414	0.6047	0.4539	0.3075	0.1836	0.0933	0.0383	0.0116	0.0022	0.0002	0.0000
	8	0.9757	0.9417	0.8811	0.7880	0.6627	0.5141	0.3595	0.2195	0.1117	0.0439	0.0115	0.0015	0.0000
	9	0.9940	0.9825	0.9574	0.9102	0.8328	0.7207	0.5773	0.4158	0.2585	0.1298	0.0467	0.0092	0.0004
	10	0.9989	0.9961	0.9886	0.9713	0.9368	0.8757	0.7795	0.6448	0.4787	0.3018	0.1465	0.0441	0.0042
	11	0.9999	0.9994	0.9978	0.9935	0.9830	0.9602	0.9161	0.8392	0.7189	0.5519	0.3521	0.1584	0.0301
	12	1.0000	0.9999	0.9997	0.9991	0.9971	0.9919	0.9795	0.9525	0.8990	0.8021	0.6433	0.4154	0.1530
	13	1.0000	1.0000	1.0000	0.9999	0.9998	0.9992	0.9976	0.9932	0.9822	0.9560	0.8972	0.7712	0.5123
15	0	0.0016	0.0005	0.0001	0.0000	0.0000	0.0000	0.0000	0.0000	0.0000	0.0000	0.0000	0.0000	0.0000
	1	0.0142	0.0052	0.0017	0.0005	0.0001	0.0000	0.0000	0.0000	0.0000	0.0000	0.0000	0.0000	0.0000
	2	0.0617	0.0271	0.0107	0.0037	0.0011	0.0003	0.0001	0.0000	0.0000	0.0000	0.0000	0.0000	0.0000
	3	0.1727	0.0905	0.0424	0.0176	0.0063	0.0019	0.0005	0.0001	0.0000	0.0000	0.0000	0.0000	0.0000
	4	0.3519	0.2173	0.1204	0.0592	0.0255	0.0093	0.0028	0.0007	0.0001	0.0000	0.0000	0.0000	0.0000
	5	0.5643	0.4032	0.2608	0.1509	0.0769	0.0338	0.0124	0.0037	0.0008	0.0001	0.0000	0.0000	0.0000
	6	0.7548	0.6098	0.4522	0.3036	0.1818	0.0950	0.0422	0.0152	0.0042	0.0008	0.0001	0.0000	0.0000
	7	0.8868	0.7869	0.6535	0.5000	0.3465	0.2131	0.1132	0.0500	0.0173	0.0042	0.0006	0.0000	0.0000
	8	0.9578	0.9050	0.8182	0.6964	0.5478	0.3902	0.2452	0.1311	0.0566	0.0181	0.0036	0.0003	0.0000
	9	0.9876	0.9662	0.9231	0.8491	0.7392	0.5968	0.4357	0.2784	0.1484	0.0611	0.0168	0.0022	0.0001
	10	0.9972	0.9907	0.9745	0.9408	0.8796	0.7827	0.6481	0.4845	0.3135	0.1642	0.0617	0.0127	0.0006
	11	0.9995	0.9981	0.9937	0.9824	0.9576	0.9095	0.8273	0.7031	0.5387	0.3518	0.1773	0.0556	0.0055
	12	0.9999	0.9997	0.9989	0.9963	0.9893	0.9729	0.9383	0.8732	0.7639	0.6020	0.3958	0.1841	0.0362
	13	1.0000	1.0000	0.9999	0.9995	0.9983	0.9948	0.9858	0.9647	0.9198	0.8329	0.6814	0.4510	0.1710
	14	1.0000	1.0000	1.0000	1.0000	0.9999	0.9995	0.9984	0.9953	0.9866	0.9648	0.9126	0.7941	0.5367

Continued

n	r	0.01	0.015	0.02	0.025	0.03	0.035	0.04	0.045	0.05	0.10	0.15	0.20	0.25	0.30
16	0	0.8515	0.7852	0.7238	0.6669	0.6143	0.5655	0.5204	0.4787	0.4401	0.2059	0.0874	0.0352	0.0134	0.0047
	1	0.9891	0.9765	0.9601	0.9405	0.9182	0.8937	0.8673	0.8396	0.8108	0.5490	0.3186	0.1671	0.0802	0.0353
	2	0.9995	0.9984	0.9963	0.9931	0.9887	0.9829	0.9758	0.9671	0.9571	0.8159	0.6042	0.3980	0.2361	0.1268
	3	1.0000	0.9999	0.9998	0.9994	0.9989	0.9981	0.9968	0.9952	0.9930	0.9444	0.8227	0.6482	0.4613	0.2969
	4	1.0000	1.0000	1.0000	1.0000	0.9999	0.9998	0.9997	0.9995	0.9991	0.9873	0.9383	0.8358	0.6865	0.5155
	5	1.0000	1.0000	1.0000	1.0000	1.0000	1.0000	1.0000	1.0000	0.9999	0.9978	0.9832	0.9389	0.8516	0.7216
	6	1.0000	1.0000	1.0000	1.0000	1.0000	1.0000	1.0000	1.0000	1.0000	0.9997	0.9964	0.9819	0.9434	0.8689
	7	1.0000	1.0000	1.0000	1.0000	1.0000	1.0000	1.0000	1.0000	1.0000	1.0000	0.9994	0.9958	0.9827	0.9500
	8	1.0000	1.0000	1.0000	1.0000	1.0000	1.0000	1.0000	1.0000	1.0000	1.0000	0.9999	0.9992	0.9958	0.9848
	9	1.0000	1.0000	1.0000	1.0000	1.0000	1.0000	1.0000	1.0000	1.0000	1.0000	1.0000	0.9999	0.9992	0.9963
	10	1.0000	1.0000	1.0000	1.0000	1.0000	1.0000	1.0000	1.0000	1.0000	1.0000	1.0000	1.0000	0.9999	0.9993
	11	1.0000	1.0000	1.0000	1.0000	1.0000	1.0000	1.0000	1.0000	1.0000	1.0000	1.0000	1.0000	1.0000	0.9999
	12	1.0000	1.0000	1.0000	1.0000	1.0000	1.0000	1.0000	1.0000	1.0000	1.0000	1.0000	1.0000	1.0000	1.0000
	13	1.0000	1.0000	1.0000	1.0000	1.0000	1.0000	1.0000	1.0000	1.0000	1.0000	1.0000	1.0000	1.0000	1.0000
	14	1.0000	1.0000	1.0000	1.0000	1.0000	1.0000	1.0000	1.0000	1.0000	1.0000	1.0000	1.0000	1.0000	1.0000
	15	1.0000	1.0000	1.0000	1.0000	1.0000	1.0000	1.0000	1.0000	1.0000	1.0000	1.0000	1.0000	1.0000	1.0000
17	0	0.8429	0.7734	0.7093	0.6502	0.5958	0.5457	0.4996	0.4571	0.4181	0.1668	0.0631	0.0225	0.0075	0.0023
	1	0.9877	0.9736	0.9554	0.9337	0.9091	0.8822	0.8535	0.8233	0.7922	0.4818	0.2525	0.1182	0.0501	0.0193
	2	0.9994	0.9980	0.9956	0.9918	0.9866	0.9798	0.9714	0.9614	0.9497	0.7618	0.5198	0.3096	0.1637	0.0774
	3	1.0000	0.9999	0.9997	0.9993	0.9986	0.9975	0.9960	0.9939	0.9912	0.9174	0.7556	0.5489	0.3530	0.2019
	4	1.0000	1.0000	1.0000	1.0000	0.9999	0.9998	0.9996	0.9993	0.9988	0.9779	0.9013	0.7582	0.5739	0.3887
	5	1.0000	1.0000	1.0000	1.0000	1.0000	1.0000	1.0000	0.9999	0.9999	0.9953	0.9681	0.8943	0.7653	0.5968
	6	1.0000	1.0000	1.0000	1.0000	1.0000	1.0000	1.0000	1.0000	1.0000	0.9992	0.9917	0.9623	0.8929	0.7752
	7	1.0000	1.0000	1.0000	1.0000	1.0000	1.0000	1.0000	1.0000	1.0000	0.9999	0.9983	0.9891	0.9598	0.8954
	8	1.0000	1.0000	1.0000	1.0000	1.0000	1.0000	1.0000	1.0000	1.0000	1.0000	0.9997	0.9974	0.9876	0.9597
	9	1.0000	1.0000	1.0000	1.0000	1.0000	1.0000	1.0000	1.0000	1.0000	1.0000	1.0000	0.9995	0.9969	0.9873
	10	1.0000	1.0000	1.0000	1.0000	1.0000	1.0000	1.0000	1.0000	1.0000	1.0000	1.0000	0.9999	0.9994	0.9968
	11	1.0000	1.0000	1.0000	1.0000	1.0000	1.0000	1.0000	1.0000	1.0000	1.0000	1.0000	1.0000	0.9999	0.9993
	12	1.0000	1.0000	1.0000	1.0000	1.0000	1.0000	1.0000	1.0000	1.0000	1.0000	1.0000	1.0000	1.0000	0.9999
	13	1.0000	1.0000	1.0000	1.0000	1.0000	1.0000	1.0000	1.0000	1.0000	1.0000	1.0000	1.0000	1.0000	1.0000
	14	1.0000	1.0000	1.0000	1.0000	1.0000	1.0000	1.0000	1.0000	1.0000	1.0000	1.0000	1.0000	1.0000	1.0000
	15	1.0000	1.0000	1.0000	1.0000	1.0000	1.0000	1.0000	1.0000	1.0000	1.0000	1.0000	1.0000	1.0000	1.0000
	16	1.0000	1.0000	1.0000	1.0000	1.0000	1.0000	1.0000	1.0000	1.0000	1.0000	1.0000	1.0000	1.0000	1.0000

Continued

n	r	0.35	0.40	0.45	0.50	0.55	0.60	0.65	0.70	0.75	0.80	0.85	0.90	0.95
16	0	0.0016	0.0005	0.0001	0.0000	0.0000	0.0000	0.0000	0.0000	0.0000	0.0000	0.0000	0.0000	0.0000
	1	0.0142	0.0052	0.0017	0.0005	0.0001	0.0000	0.0000	0.0000	0.0000	0.0000	0.0000	0.0000	0.0000
	2	0.0617	0.0271	0.0107	0.0037	0.0011	0.0003	0.0001	0.0000	0.0000	0.0000	0.0000	0.0000	0.0000
	3	0.1727	0.0905	0.0424	0.0176	0.0063	0.0019	0.0005	0.0001	0.0000	0.0000	0.0000	0.0000	0.0000
	4	0.3519	0.2173	0.1204	0.0592	0.0255	0.0093	0.0028	0.0007	0.0001	0.0000	0.0000	0.0000	0.0000
	5	0.5643	0.4032	0.2608	0.1509	0.0769	0.0338	0.0124	0.0037	0.0008	0.0001	0.0000	0.0000	0.0000
	6	0.7548	0.6098	0.4522	0.3036	0.1818	0.0950	0.0422	0.0152	0.0042	0.0008	0.0001	0.0000	0.0000
	7	0.8868	0.7869	0.6535	0.5000	0.3465	0.2131	0.1132	0.0500	0.0173	0.0042	0.0006	0.0000	0.0000
	8	0.9578	0.9050	0.8182	0.6964	0.5478	0.3902	0.2452	0.1311	0.0566	0.0181	0.0036	0.0003	0.0000
	9	0.9876	0.9662	0.9231	0.8491	0.7392	0.5968	0.4357	0.2784	0.1484	0.0611	0.0168	0.0022	0.0001
	10	0.9972	0.9907	0.9745	0.9408	0.8796	0.7827	0.6481	0.4845	0.3135	0.1642	0.0617	0.0127	0.0006
	11	0.9995	0.9981	0.9937	0.9824	0.9576	0.9095	0.8273	0.7031	0.5387	0.3518	0.1773	0.0556	0.0055
	12	0.9999	0.9997	0.9989	0.9963	0.9893	0.9729	0.9383	0.8732	0.7639	0.6020	0.3958	0.1841	0.0362
	13	1.0000	1.0000	0.9999	0.9995	0.9983	0.9948	0.9858	0.9647	0.9198	0.8329	0.6814	0.4510	0.1710
	14	1.0000	1.0000	1.0000	1.0000	0.9999	0.9995	0.9984	0.9953	0.9866	0.9648	0.9126	0.7941	0.5367
	15	1.0000	1.0000	1.0000	1.0000	1.0000	1.0000	1.0000	1.0000	1.0000	1.0000	1.0000	1.0000	1.0000
17	0	0.0007	0.0002	0.0000	0.0000	0.0000	0.0000	0.0000	0.0000	0.0000	0.0000	0.0000	0.0000	0.0000
	1	0.0067	0.0021	0.0006	0.0001	0.0000	0.0000	0.0000	0.0000	0.0000	0.0000	0.0000	0.0000	0.0000
	2	0.0327	0.0123	0.0041	0.0012	0.0003	0.0001	0.0000	0.0000	0.0000	0.0000	0.0000	0.0000	0.0000
	3	0.1028	0.0464	0.0184	0.0064	0.0019	0.0005	0.0001	0.0000	0.0000	0.0000	0.0000	0.0000	0.0000
	4	0.2348	0.1260	0.0596	0.0245	0.0086	0.0025	0.0006	0.0001	0.0000	0.0000	0.0000	0.0000	0.0000
	5	0.4197	0.2639	0.1471	0.0717	0.0301	0.0106	0.0030	0.0007	0.0001	0.0000	0.0000	0.0000	0.0000
	6	0.6188	0.4478	0.2902	0.1662	0.0826	0.0348	0.0120	0.0032	0.0006	0.0001	0.0000	0.0000	0.0000
	7	0.7872	0.6405	0.4743	0.3145	0.1834	0.0919	0.0383	0.0127	0.0031	0.0005	0.0000	0.0000	0.0000
	8	0.9006	0.8011	0.6626	0.5000	0.3374	0.1989	0.0994	0.0403	0.0124	0.0026	0.0003	0.0000	0.0000
	9	0.9617	0.9081	0.8166	0.6855	0.5257	0.3595	0.2128	0.1046	0.0402	0.0109	0.0017	0.0001	0.0000
	10	0.9880	0.9652	0.9174	0.8338	0.7098	0.5522	0.3812	0.2248	0.1071	0.0377	0.0083	0.0008	0.0000
	11	0.9970	0.9894	0.9699	0.9283	0.8529	0.7361	0.5803	0.4032	0.2347	0.1057	0.0319	0.0047	0.0001
	12	0.9994	0.9975	0.9914	0.9755	0.9404	0.8740	0.7652	0.6113	0.4261	0.2418	0.0987	0.0221	0.0012
	13	0.9999	0.9995	0.9981	0.9936	0.9816	0.9536	0.8972	0.7981	0.6470	0.4511	0.2444	0.0826	0.0088
	14	1.0000	0.9999	0.9997	0.9988	0.9959	0.9877	0.9673	0.9226	0.8363	0.6904	0.4802	0.2382	0.0503
	15	1.0000	1.0000	1.0000	0.9999	0.9994	0.9979	0.9933	0.9807	0.9499	0.8818	0.7475	0.5182	0.2078
	16	1.0000	1.0000	1.0000	1.0000	1.0000	0.9998	0.9993	0.9977	0.9925	0.9775	0.9369	0.8332	0.5819

Continued

n	r	0.01	0.015	0.02	0.025	0.03	0.035	0.04	0.045	0.05	0.10	0.15	0.20	0.25	0.30
18	0	0.8345	0.7618	0.6951	0.6340	0.5780	0.5266	0.4796	0.4366	0.3972	0.1501	0.0536	0.0180	0.0056	0.0016
	1	0.9862	0.9706	0.9505	0.9266	0.8997	0.8704	0.8393	0.8069	0.7735	0.4503	0.2241	0.0991	0.0395	0.0142
	2	0.9993	0.9977	0.9948	0.9904	0.9843	0.9764	0.9667	0.9552	0.9419	0.7338	0.4797	0.2713	0.1353	0.0600
	3	1.0000	0.9999	0.9996	0.9991	0.9982	0.9969	0.9950	0.9924	0.9891	0.9018	0.7202	0.5010	0.3057	0.1646
	4	1.0000	1.0000	1.0000	0.9999	0.9998	0.9997	0.9994	0.9990	0.9985	0.9718	0.8794	0.7164	0.5187	0.3327
	5	1.0000	1.0000	1.0000	1.0000	1.0000	1.0000	0.9999	0.9999	0.9998	0.9936	0.9581	0.8671	0.7175	0.5344
	6	1.0000	1.0000	1.0000	1.0000	1.0000	1.0000	1.0000	1.0000	1.0000	0.9988	0.9882	0.9487	0.8610	0.7217
	7	1.0000	1.0000	1.0000	1.0000	1.0000	1.0000	1.0000	1.0000	1.0000	0.9998	0.9973	0.9837	0.9431	0.8593
	8	1.0000	1.0000	1.0000	1.0000	1.0000	1.0000	1.0000	1.0000	1.0000	1.0000	0.9995	0.9957	0.9807	0.9404
	9	1.0000	1.0000	1.0000	1.0000	1.0000	1.0000	1.0000	1.0000	1.0000	1.0000	0.9999	0.9991	0.9946	0.9790
	10	1.0000	1.0000	1.0000	1.0000	1.0000	1.0000	1.0000	1.0000	1.0000	1.0000	1.0000	0.9998	0.9988	0.9939
	11	1.0000	1.0000	1.0000	1.0000	1.0000	1.0000	1.0000	1.0000	1.0000	1.0000	1.0000	1.0000	0.9998	0.9986
	12	1.0000	1.0000	1.0000	1.0000	1.0000	1.0000	1.0000	1.0000	1.0000	1.0000	1.0000	1.0000	1.0000	0.9997
	13	1.0000	1.0000	1.0000	1.0000	1.0000	1.0000	1.0000	1.0000	1.0000	1.0000	1.0000	1.0000	1.0000	1.0000
	14	1.0000	1.0000	1.0000	1.0000	1.0000	1.0000	1.0000	1.0000	1.0000	1.0000	1.0000	1.0000	1.0000	1.0000
	15	1.0000	1.0000	1.0000	1.0000	1.0000	1.0000	1.0000	1.0000	1.0000	1.0000	1.0000	1.0000	1.0000	1.0000
	16	1.0000	1.0000	1.0000	1.0000	1.0000	1.0000	1.0000	1.0000	1.0000	1.0000	1.0000	1.0000	1.0000	1.0000
	17	1.0000	1.0000	1.0000	1.0000	1.0000	1.0000	1.0000	1.0000	1.0000	1.0000	1.0000	1.0000	1.0000	1.0000
19	0	0.8262	0.7504	0.6812	0.6181	0.5606	0.5082	0.4604	0.4169	0.3774	0.1351	0.0456	0.0144	0.0042	0.0011
	1	0.9847	0.9675	0.9454	0.9193	0.8900	0.8584	0.8249	0.7902	0.7547	0.4203	0.1985	0.0829	0.0310	0.0104
	2	0.9991	0.9973	0.9939	0.9888	0.9817	0.9727	0.9616	0.9485	0.9335	0.7054	0.4413	0.2369	0.1113	0.0462
	3	1.0000	0.9998	0.9995	0.9989	0.9978	0.9962	0.9939	0.9908	0.9868	0.8850	0.6841	0.4551	0.2631	0.1332
	4	1.0000	1.0000	1.0000	0.9999	0.9998	0.9996	0.9993	0.9987	0.9980	0.9648	0.8556	0.6733	0.4654	0.2822
	5	1.0000	1.0000	1.0000	1.0000	1.0000	1.0000	0.9999	0.9999	0.9998	0.9914	0.9463	0.8369	0.6678	0.4739
	6	1.0000	1.0000	1.0000	1.0000	1.0000	1.0000	1.0000	1.0000	1.0000	0.9983	0.9837	0.9324	0.8251	0.6655
	7	1.0000	1.0000	1.0000	1.0000	1.0000	1.0000	1.0000	1.0000	1.0000	0.9997	0.9959	0.9767	0.9225	0.8180
	8	1.0000	1.0000	1.0000	1.0000	1.0000	1.0000	1.0000	1.0000	1.0000	1.0000	0.9992	0.9933	0.9713	0.9161
	9	1.0000	1.0000	1.0000	1.0000	1.0000	1.0000	1.0000	1.0000	1.0000	1.0000	0.9999	0.9984	0.9911	0.9674
	10	1.0000	1.0000	1.0000	1.0000	1.0000	1.0000	1.0000	1.0000	1.0000	1.0000	1.0000	0.9997	0.9977	0.9895
	11	1.0000	1.0000	1.0000	1.0000	1.0000	1.0000	1.0000	1.0000	1.0000	1.0000	1.0000	1.0000	0.9995	0.9972
	12	1.0000	1.0000	1.0000	1.0000	1.0000	1.0000	1.0000	1.0000	1.0000	1.0000	1.0000	1.0000	0.9999	0.9994
	13	1.0000	1.0000	1.0000	1.0000	1.0000	1.0000	1.0000	1.0000	1.0000	1.0000	1.0000	1.0000	1.0000	0.9999
	14	1.0000	1.0000	1.0000	1.0000	1.0000	1.0000	1.0000	1.0000	1.0000	1.0000	1.0000	1.0000	1.0000	1.0000
	15	1.0000	1.0000	1.0000	1.0000	1.0000	1.0000	1.0000	1.0000	1.0000	1.0000	1.0000	1.0000	1.0000	1.0000
	16	1.0000	1.0000	1.0000	1.0000	1.0000	1.0000	1.0000	1.0000	1.0000	1.0000	1.0000	1.0000	1.0000	1.0000
	17	1.0000	1.0000	1.0000	1.0000	1.0000	1.0000	1.0000	1.0000	1.0000	1.0000	1.0000	1.0000	1.0000	1.0000
	18	1.0000	1.0000	1.0000	1.0000	1.0000	1.0000	1.0000	1.0000	1.0000	1.0000	1.0000	1.0000	1.0000	1.0000

Continued

n	r	0.35	0.40	0.45	0.50	0.55	0.60	0.65	0.70	0.75	0.80	0.85	0.90	0.95
18	0	0.0004	0.0001	0.0000	0.0000	0.0000	0.0000	0.0000	0.0000	0.0000	0.0000	0.0000	0.0000	0.0000
	1	0.0046	0.0013	0.0003	0.0001	0.0000	0.0000	0.0000	0.0000	0.0000	0.0000	0.0000	0.0000	0.0000
	2	0.0236	0.0082	0.0025	0.0007	0.0001	0.0000	0.0000	0.0000	0.0000	0.0000	0.0000	0.0000	0.0000
	3	0.0783	0.0328	0.0120	0.0038	0.0010	0.0002	0.0000	0.0000	0.0000	0.0000	0.0000	0.0000	0.0000
	4	0.1886	0.0942	0.0411	0.0154	0.0049	0.0013	0.0003	0.0000	0.0000	0.0000	0.0000	0.0000	0.0000
	5	0.3550	0.2088	0.1077	0.0481	0.0183	0.0058	0.0014	0.0003	0.0000	0.0000	0.0000	0.0000	0.0000
	6	0.5491	0.3743	0.2258	0.1189	0.0537	0.0203	0.0062	0.0014	0.0002	0.0000	0.0000	0.0000	0.0000
	7	0.7283	0.5634	0.3915	0.2403	0.1280	0.0576	0.0212	0.0061	0.0012	0.0002	0.0000	0.0000	0.0000
	8	0.8609	0.7368	0.5778	0.4073	0.2527	0.1347	0.0597	0.0210	0.0054	0.0009	0.0001	0.0000	0.0000
	9	0.9403	0.8653	0.7473	0.5927	0.4222	0.2632	0.1391	0.0596	0.0193	0.0043	0.0005	0.0000	0.0000
	10	0.9788	0.9424	0.8720	0.7597	0.6085	0.4366	0.2717	0.1407	0.0569	0.0163	0.0027	0.0002	0.0000
	11	0.9938	0.9797	0.9463	0.8811	0.7742	0.6257	0.4509	0.2783	0.1390	0.0513	0.0118	0.0012	0.0000
	12	0.9986	0.9942	0.9817	0.9519	0.8923	0.7912	0.6450	0.4656	0.2825	0.1329	0.0419	0.0064	0.0002
	13	0.9997	0.9987	0.9951	0.9846	0.9589	0.9058	0.8114	0.6673	0.4813	0.2836	0.1206	0.0282	0.0015
	14	1.0000	0.9998	0.9990	0.9962	0.9880	0.9672	0.9217	0.8354	0.6943	0.4990	0.2798	0.0982	0.0109
	15	1.0000	1.0000	0.9999	0.9993	0.9975	0.9918	0.9764	0.9400	0.8647	0.7287	0.5203	0.2662	0.0581
	16	1.0000	1.0000	1.0000	0.9999	0.9997	0.9987	0.9954	0.9858	0.9605	0.9009	0.7759	0.5497	0.2265
	17	1.0000	1.0000	1.0000	1.0000	1.0000	0.9999	0.9996	0.9984	0.9944	0.9820	0.9464	0.8499	0.6028
19	0	0.0003	0.0001	0.0000	0.0000	0.0000	0.0000	0.0000	0.0000	0.0000	0.0000	0.0000	0.0000	0.0000
	1	0.0031	0.0008	0.0002	0.0000	0.0000	0.0000	0.0000	0.0000	0.0000	0.0000	0.0000	0.0000	0.0000
	2	0.0170	0.0055	0.0015	0.0004	0.0001	0.0000	0.0000	0.0000	0.0000	0.0000	0.0000	0.0000	0.0000
	3	0.0591	0.0230	0.0077	0.0022	0.0005	0.0001	0.0000	0.0000	0.0000	0.0000	0.0000	0.0000	0.0000
	4	0.1500	0.0696	0.0280	0.0096	0.0028	0.0006	0.0001	0.0000	0.0000	0.0000	0.0000	0.0000	0.0000
	5	0.2968	0.1629	0.0777	0.0318	0.0109	0.0031	0.0007	0.0001	0.0000	0.0000	0.0000	0.0000	0.0000
	6	0.4812	0.3081	0.1727	0.0835	0.0342	0.0116	0.0031	0.0006	0.0001	0.0000	0.0000	0.0000	0.0000
	7	0.6656	0.4878	0.3169	0.1796	0.0871	0.0352	0.0114	0.0028	0.0005	0.0000	0.0000	0.0000	0.0000
	8	0.8145	0.6675	0.4940	0.3238	0.1841	0.0885	0.0347	0.0105	0.0023	0.0003	0.0000	0.0000	0.0000
	9	0.9125	0.8139	0.6710	0.5000	0.3290	0.1861	0.0875	0.0326	0.0089	0.0016	0.0001	0.0000	0.0000
	10	0.9653	0.9115	0.8159	0.6762	0.5060	0.3325	0.1855	0.0839	0.0287	0.0067	0.0008	0.0000	0.0000
	11	0.9886	0.9648	0.9129	0.8204	0.6831	0.5122	0.3344	0.1820	0.0775	0.0233	0.0041	0.0003	0.0000
	12	0.9969	0.9884	0.9658	0.9165	0.8273	0.6919	0.5188	0.3345	0.1749	0.0676	0.0163	0.0017	0.0000
	13	0.9993	0.9969	0.9891	0.9682	0.9223	0.8371	0.7032	0.5261	0.3322	0.1631	0.0537	0.0086	0.0002
	14	0.9999	0.9994	0.9972	0.9904	0.9720	0.9304	0.8500	0.7178	0.5346	0.3267	0.1444	0.0352	0.0020
	15	1.0000	0.9999	0.9995	0.9978	0.9923	0.9770	0.9409	0.8668	0.7369	0.5449	0.3159	0.1150	0.0132
	16	1.0000	1.0000	0.9999	0.9996	0.9985	0.9945	0.9830	0.9538	0.8887	0.7631	0.5587	0.2946	0.0665
	17	1.0000	1.0000	1.0000	1.0000	0.9998	0.9992	0.9969	0.9896	0.9690	0.9171	0.8015	0.5797	0.2453
	18	1.0000	1.0000	1.0000	1.0000	1.0000	0.9999	0.9997	0.9989	0.9958	0.9856	0.9544	0.8649	0.6226

Continued

n	r	0.01	0.015	0.02	0.025	0.03	0.035	0.04	0.045	0.05	0.10	0.15	0.20	0.25	0.30
20	0	0.8179	0.7391	0.6676	0.6027	0.5438	0.4904	0.4420	0.3982	0.3585	0.1216	0.0388	0.0115	0.0032	0.0008
	1	0.9831	0.9643	0.9401	0.9118	0.8802	0.8461	0.8103	0.7734	0.7358	0.3917	0.1756	0.0692	0.0243	0.0076
	2	0.9990	0.9968	0.9929	0.9870	0.9790	0.9687	0.9561	0.9414	0.9245	0.6769	0.4049	0.2061	0.0913	0.0355
	3	1.0000	0.9998	0.9994	0.9986	0.9973	0.9954	0.9926	0.9889	0.9841	0.8670	0.6477	0.4114	0.2252	0.1071
	4	1.0000	1.0000	1.0000	0.9999	0.9997	0.9995	0.9990	0.9984	0.9974	0.9568	0.8298	0.6296	0.4148	0.2375
	5	1.0000	1.0000	1.0000	1.0000	1.0000	1.0000	0.9999	0.9998	0.9997	0.9887	0.9327	0.8042	0.6172	0.4164
	6	1.0000	1.0000	1.0000	1.0000	1.0000	1.0000	1.0000	1.0000	1.0000	0.9976	0.9781	0.9133	0.7858	0.6080
	7	1.0000	1.0000	1.0000	1.0000	1.0000	1.0000	1.0000	1.0000	1.0000	0.9996	0.9941	0.9679	0.8982	0.7723
	8	1.0000	1.0000	1.0000	1.0000	1.0000	1.0000	1.0000	1.0000	1.0000	0.9999	0.9987	0.9900	0.9591	0.8867
	9	1.0000	1.0000	1.0000	1.0000	1.0000	1.0000	1.0000	1.0000	1.0000	1.0000	0.9998	0.9974	0.9861	0.9520
	10	1.0000	1.0000	1.0000	1.0000	1.0000	1.0000	1.0000	1.0000	1.0000	1.0000	1.0000	0.9994	0.9961	0.9829
	11	1.0000	1.0000	1.0000	1.0000	1.0000	1.0000	1.0000	1.0000	1.0000	1.0000	1.0000	0.9999	0.9991	0.9949
	12	1.0000	1.0000	1.0000	1.0000	1.0000	1.0000	1.0000	1.0000	1.0000	1.0000	1.0000	1.0000	0.9998	0.9987
	13	1.0000	1.0000	1.0000	1.0000	1.0000	1.0000	1.0000	1.0000	1.0000	1.0000	1.0000	1.0000	1.0000	0.9997
	14	1.0000	1.0000	1.0000	1.0000	1.0000	1.0000	1.0000	1.0000	1.0000	1.0000	1.0000	1.0000	1.0000	1.0000
	15	1.0000	1.0000	1.0000	1.0000	1.0000	1.0000	1.0000	1.0000	1.0000	1.0000	1.0000	1.0000	1.0000	1.0000
	16	1.0000	1.0000	1.0000	1.0000	1.0000	1.0000	1.0000	1.0000	1.0000	1.0000	1.0000	1.0000	1.0000	1.0000
	17	1.0000	1.0000	1.0000	1.0000	1.0000	1.0000	1.0000	1.0000	1.0000	1.0000	1.0000	1.0000	1.0000	1.0000
	18	1.0000	1.0000	1.0000	1.0000	1.0000	1.0000	1.0000	1.0000	1.0000	1.0000	1.0000	1.0000	1.0000	1.0000
	19	1.0000	1.0000	1.0000	1.0000	1.0000	1.0000	1.0000	1.0000	1.0000	1.0000	1.0000	1.0000	1.0000	1.0000

n	*r*	0.35	0.40	0.45	0.50	0.55	0.60	0.65	0.70	0.75	0.80	0.85	0.90	0.95
20	0	0.0002	0.0000	0.0000	0.0000	0.0000	0.0000	0.0000	0.0000	0.0000	0.0000	0.0000	0.0000	0.0000
	1	0.0021	0.0005	0.0001	0.0000	0.0000	0.0000	0.0000	0.0000	0.0000	0.0000	0.0000	0.0000	0.0000
	2	0.0121	0.0036	0.0009	0.0002	0.0000	0.0000	0.0000	0.0000	0.0000	0.0000	0.0000	0.0000	0.0000
	3	0.0444	0.0160	0.0049	0.0013	0.0003	0.0000	0.0000	0.0000	0.0000	0.0000	0.0000	0.0000	0.0000
	4	0.1182	0.0510	0.0189	0.0059	0.0015	0.0003	0.0000	0.0000	0.0000	0.0000	0.0000	0.0000	0.0000
	5	0.2454	0.1256	0.0553	0.0207	0.0064	0.0016	0.0003	0.0000	0.0000	0.0000	0.0000	0.0000	0.0000
	6	0.4166	0.2500	0.1299	0.0577	0.0214	0.0065	0.0015	0.0003	0.0000	0.0000	0.0000	0.0000	0.0000
	7	0.6010	0.4159	0.2520	0.1316	0.0580	0.0210	0.0060	0.0013	0.0002	0.0000	0.0000	0.0000	0.0000
	8	0.7624	0.5956	0.4143	0.2517	0.1308	0.0565	0.0196	0.0051	0.0009	0.0001	0.0000	0.0000	0.0000
	9	0.8782	0.7553	0.5914	0.4119	0.2493	0.1275	0.0532	0.0171	0.0039	0.0006	0.0000	0.0000	0.0000
	10	0.9468	0.8725	0.7507	0.5881	0.4086	0.2447	0.1218	0.0480	0.0139	0.0026	0.0002	0.0000	0.0000
	11	0.9804	0.9435	0.8692	0.7483	0.5857	0.4044	0.2376	0.1133	0.0409	0.0100	0.0013	0.0001	0.0000
	12	0.9940	0.9790	0.9420	0.8684	0.7480	0.5841	0.3990	0.2277	0.1018	0.0321	0.0059	0.0004	0.0000
	13	0.9985	0.9935	0.9786	0.9423	0.8701	0.7500	0.5834	0.3920	0.2142	0.0867	0.0219	0.0024	0.0000
	14	0.9997	0.9984	0.9936	0.9793	0.9447	0.8744	0.7546	0.5836	0.3828	0.1958	0.0673	0.0113	0.0003
	15	1.0000	0.9997	0.9985	0.9941	0.9811	0.9490	0.8818	0.7625	0.5852	0.3704	0.1702	0.0432	0.0026
	16	1.0000	1.0000	0.9997	0.9987	0.9951	0.9840	0.9556	0.8929	0.7748	0.5886	0.3523	0.1330	0.0159
	17	1.0000	1.0000	1.0000	0.9998	0.9991	0.9964	0.9879	0.9645	0.9087	0.7939	0.5951	0.3231	0.0755
	18	1.0000	1.0000	1.0000	1.0000	0.9999	0.9995	0.9979	0.9924	0.9757	0.9308	0.8244	0.6083	0.2642
	19	1.0000	1.0000	1.0000	1.0000	1.0000	1.0000	0.9998	0.9992	0.9968	0.9885	0.9612	0.8784	0.6415

Appendix N
The Poisson Distribution

np \ r	0	1	2	3	4	5	6	7
0.01	0.990							
0.02	0.980							
0.03	0.970	1.000						
0.04	0.961	0.999						
0.05	0.951	0.999						
0.06	0.942	0.998						
0.07	0.932	0.998						
0.08	0.923	0.997						
0.09	0.914	0.996						
0.10	0.905	0.995	1.000					
0.15	0.861	0.990	0.999					
0.20	0.819	0.982	0.999					
0.25	0.779	0.974	0.998					
0.30	0.741	0.963	0.996					
0.35	0.705	0.951	0.994	1.000				
0.40	0.670	0.938	0.992	0.999				
0.45	0.638	0.925	0.989	0.999				
0.50	0.607	0.910	0.986	0.998				
0.60	0.549	0.878	0.977	0.997	1.000			
0.65	0.522	0.861	0.972	0.996	0.999			
0.70	0.497	0.844	0.966	0.994	0.999			
0.75	0.472	0.827	0.959	0.993	0.999			
0.80	0.449	0.809	0.953	0.991	0.999			
0.85	0.427	0.791	0.945	0.989	0.998			
0.90	0.407	0.772	0.937	0.987	0.998			
0.95	0.387	0.754	0.929	0.984	0.997	1.000		
1.00	0.368	0.736	0.920	0.981	0.996	0.999		
1.10	0.333	0.699	0.900	0.974	0.995	0.999		
1.20	0.301	0.663	0.879	0.966	0.992	0.998		
1.30	0.273	0.627	0.857	0.957	0.989	0.998	1.000	
1.40	0.247	0.592	0.833	0.946	0.986	0.997	0.999	
1.50	0.223	0.558	0.809	0.934	0.981	0.996	0.999	
1.60	0.202	0.525	0.783	0.921	0.976	0.994	0.999	
1.70	0.183	0.493	0.757	0.907	0.970	0.992	0.998	1.000

Continued

np \ r	0	1	2	3	4	5	6	7	8	9	10	11	12	13	14	15	16	17	18
1.80	0.165	0.463	0.731	0.891	0.964	0.990	0.997	0.999											
1.90	0.150	0.434	0.704	0.875	0.956	0.987	0.997	0.999											
2.00	0.135	0.406	0.677	0.857	0.947	0.983	0.995	0.999											
2.20	0.111	0.355	0.623	0.819	0.928	0.975	0.993	0.998	1.000										
2.40	0.091	0.308	0.570	0.779	0.904	0.964	0.988	0.997	0.999										
2.60	0.074	0.267	0.518	0.736	0.877	0.951	0.983	0.995	0.999	1.000									
2.80	0.061	0.231	0.469	0.692	0.848	0.935	0.976	0.992	0.998	0.999									
3.00	0.050	0.199	0.423	0.647	0.815	0.916	0.966	0.988	0.996	0.999									
3.20	0.041	0.171	0.380	0.603	0.781	0.895	0.955	0.983	0.994	0.998	1.000								
3.40	0.033	0.147	0.340	0.558	0.744	0.871	0.942	0.977	0.992	0.997	0.999								
3.60	0.027	0.126	0.303	0.515	0.706	0.844	0.927	0.969	0.988	0.996	0.999	1.000							
3.80	0.022	0.107	0.269	0.473	0.668	0.816	0.909	0.960	0.984	0.994	0.998	0.999							
4.00	0.018	0.092	0.238	0.433	0.629	0.785	0.889	0.949	0.979	0.992	0.997	0.999							
4.20	0.015	0.078	0.210	0.395	0.590	0.753	0.867	0.936	0.972	0.989	0.996	0.999	1.000						
4.40	0.012	0.066	0.185	0.359	0.551	0.720	0.844	0.921	0.964	0.985	0.994	0.998	0.999						
4.60	0.010	0.056	0.163	0.326	0.513	0.686	0.818	0.905	0.955	0.980	0.992	0.997	0.999						
4.80	0.008	0.048	0.143	0.294	0.476	0.651	0.791	0.887	0.944	0.975	0.990	0.996	0.999	1.000					
5.00	0.007	0.040	0.125	0.265	0.440	0.616	0.762	0.867	0.932	0.968	0.986	0.995	0.998	0.999					
5.20	0.006	0.034	0.109	0.238	0.406	0.581	0.732	0.845	0.918	0.960	0.982	0.993	0.997	0.999					
5.40	0.005	0.029	0.095	0.213	0.373	0.546	0.702	0.822	0.903	0.951	0.977	0.990	0.996	0.999	1.000				
5.60	0.004	0.024	0.082	0.191	0.342	0.512	0.670	0.797	0.886	0.941	0.972	0.988	0.995	0.998	0.999				
5.80	0.003	0.021	0.072	0.170	0.313	0.478	0.638	0.771	0.867	0.929	0.965	0.984	0.993	0.997	0.999	1.000			
6.00	0.002	0.017	0.062	0.151	0.285	0.446	0.606	0.744	0.847	0.916	0.957	0.980	0.991	0.996	0.999	0.999			
6.20	0.002	0.015	0.054	0.134	0.259	0.414	0.574	0.716	0.826	0.902	0.949	0.975	0.989	0.995	0.998	0.999			
6.40	0.002	0.012	0.046	0.119	0.235	0.384	0.542	0.687	0.803	0.886	0.939	0.969	0.986	0.994	0.997	0.999	1.000		
6.60	0.001	0.010	0.040	0.105	0.213	0.355	0.511	0.658	0.780	0.869	0.927	0.963	0.982	0.992	0.997	0.999	0.999		
6.80	0.001	0.009	0.034	0.093	0.192	0.327	0.480	0.628	0.755	0.850	0.915	0.955	0.978	0.990	0.996	0.998	0.999		
7.00	0.001	0.007	0.030	0.082	0.173	0.301	0.450	0.599	0.729	0.830	0.901	0.947	0.973	0.987	0.994	0.998	0.999		
7.20	0.001	0.006	0.025	0.072	0.156	0.276	0.420	0.569	0.703	0.810	0.887	0.937	0.967	0.984	0.993	0.997	0.999	1.000	
7.40	0.001	0.005	0.022	0.063	0.140	0.253	0.392	0.539	0.676	0.788	0.871	0.926	0.961	0.980	0.991	0.996	0.998	0.999	
7.60	0.001	0.004	0.019	0.055	0.125	0.231	0.365	0.510	0.648	0.765	0.854	0.915	0.954	0.976	0.989	0.995	0.998	0.999	
7.80	0.000	0.004	0.016	0.048	0.112	0.210	0.338	0.481	0.620	0.741	0.835	0.902	0.945	0.971	0.986	0.993	0.997	0.999	1.000
8.00	0.000	0.003	0.014	0.042	0.100	0.191	0.313	0.453	0.593	0.717	0.816	0.888	0.936	0.966	0.983	0.992	0.996	0.998	0.999
8.20	0.000	0.003	0.012	0.037	0.089	0.174	0.290	0.425	0.565	0.692	0.796	0.873	0.926	0.960	0.979	0.990	0.995	0.998	0.999

Continued

np \ r	0	1	2	3	4	5	6	7	8	9	10	11	12	13	14	15	16	17	18
8.40	0.000	0.002	0.010	0.032	0.079	0.157	0.267	0.399	0.537	0.666	0.774	0.857	0.915	0.952	0.975	0.987	0.994	0.997	0.999
8.60	0.000	0.002	0.009	0.028	0.070	0.142	0.246	0.373	0.509	0.640	0.752	0.840	0.903	0.945	0.970	0.985	0.993	0.997	0.999
8.80	0.000	0.001	0.007	0.024	0.062	0.128	0.226	0.348	0.482	0.614	0.729	0.822	0.890	0.936	0.965	0.982	0.991	0.996	0.998
9.00	0.000	0.001	0.006	0.021	0.055	0.116	0.207	0.324	0.456	0.587	0.706	0.803	0.876	0.926	0.959	0.978	0.989	0.995	0.998
9.20	0.000	0.001	0.005	0.018	0.049	0.104	0.189	0.301	0.430	0.561	0.682	0.783	0.861	0.916	0.952	0.974	0.987	0.993	0.997
9.40	0.000	0.001	0.005	0.016	0.043	0.093	0.173	0.279	0.404	0.535	0.658	0.763	0.845	0.904	0.944	0.969	0.984	0.992	0.996
9.60	0.000	0.001	0.004	0.014	0.038	0.084	0.157	0.258	0.380	0.509	0.633	0.741	0.828	0.892	0.936	0.964	0.981	0.990	0.995
9.80	0.000	0.001	0.003	0.012	0.033	0.075	0.143	0.239	0.356	0.483	0.608	0.719	0.810	0.879	0.927	0.958	0.977	0.988	0.994
10.00	0.000	0.000	0.003	0.010	0.029	0.067	0.130	0.220	0.333	0.458	0.583	0.697	0.792	0.864	0.917	0.951	0.973	0.986	0.993
11.00	0.000	0.000	0.001	0.005	0.015	0.038	0.079	0.143	0.232	0.341	0.460	0.579	0.689	0.781	0.854	0.907	0.944	0.968	0.982
12.00	0.000	0.000	0.001	0.002	0.008	0.020	0.046	0.090	0.155	0.242	0.347	0.462	0.576	0.682	0.772	0.844	0.899	0.937	0.963
13.00	0.000	0.000	0.000	0.001	0.004	0.011	0.026	0.054	0.100	0.166	0.252	0.353	0.463	0.573	0.675	0.764	0.835	0.890	0.930
14.00	0.000	0.000	0.000	0.000	0.002	0.006	0.014	0.032	0.062	0.109	0.176	0.260	0.358	0.464	0.570	0.669	0.756	0.827	0.883
15.00	0.000	0.000	0.000	0.000	0.001	0.003	0.008	0.018	0.037	0.070	0.118	0.185	0.268	0.363	0.466	0.568	0.664	0.749	0.819
16.00	0.000	0.000	0.000	0.000	0.000	0.001	0.004	0.010	0.022	0.043	0.077	0.127	0.193	0.275	0.368	0.467	0.566	0.659	0.742
17.00	0.000	0.000	0.000	0.000	0.000	0.001	0.002	0.005	0.013	0.026	0.049	0.085	0.135	0.201	0.281	0.371	0.468	0.564	0.655
18.00	0.000	0.000	0.000	0.000	0.000	0.000	0.001	0.003	0.007	0.015	0.030	0.055	0.092	0.143	0.208	0.287	0.375	0.469	0.562
19.00	0.000	0.000	0.000	0.000	0.000	0.000	0.001	0.002	0.004	0.009	0.018	0.035	0.061	0.098	0.150	0.215	0.292	0.378	0.469
20.00	0.000	0.000	0.000	0.000	0.000	0.000	0.000	0.001	0.002	0.005	0.011	0.021	0.039	0.066	0.105	0.157	0.221	0.297	0.381

Continued

np \ r	19	20	21	22	23	24	25	26	27	28	29	30	31	32	33	34	35	36
8.40	1.000																	
8.60	0.999																	
8.80	0.999																	
9.00	0.999	1.000																
9.20	0.999	0.999																
9.40	0.998	0.999																
9.60	0.998	0.999	1.000															
9.80	0.997	0.999	0.999															
10.00	0.997	0.998	0.999	1.000														
11.00	0.991	0.995	0.998	0.999	1.000	1.000												
12.00	0.979	0.988	0.994	0.997	0.999	0.999	1.000											
13.00	0.957	0.975	0.986	0.992	0.996	0.998	0.999	1.000	1.000									
14.00	0.923	0.952	0.971	0.983	0.991	0.995	0.997	0.999	0.999	1.000								
15.00	0.875	0.917	0.947	0.967	0.981	0.989	0.994	0.997	0.998	0.999	1.000	1.000						
16.00	0.812	0.868	0.911	0.942	0.963	0.978	0.987	0.993	0.996	0.998	0.999	0.999	1.000					
17.00	0.736	0.805	0.861	0.905	0.937	0.959	0.975	0.985	0.991	0.995	0.997	0.999	0.999	1.000				
18.00	0.651	0.731	0.799	0.855	0.899	0.932	0.955	0.972	0.983	0.990	0.994	0.997	0.998	0.999	1.000	1.000		
19.00	0.561	0.647	0.725	0.793	0.849	0.893	0.927	0.951	0.969	0.980	0.988	0.993	0.996	0.998	0.999	0.999	1.000	
20.00	0.470	0.559	0.644	0.721	0.787	0.843	0.888	0.922	0.948	0.966	0.978	0.987	0.992	0.995	0.997	0.999	0.999	1.000

Appendix O
Control Chart Constants

(n)	A	A_2	D_1	D_2	D_3	D_4	A_3	B_3	B_4	d_2	d_3	c_4
2	2.121	1.880	0.000	3.686	0.000	3.267	2.659	0.000	3.267	1.128	0.853	0.798
3	1.732	1.023	0.000	4.358	0.000	2.574	1.954	0.000	2.568	1.693	0.888	0.886
4	1.500	0.729	0.000	4.698	0.000	2.282	1.628	0.000	2.266	2.059	0.880	0.921
5	1.342	0.577	0.000	4.918	0.000	2.114	1.427	0.000	2.089	2.326	0.864	0.940
6	1.225	0.483	0.000	5.078	0.000	2.004	1.287	0.030	1.970	2.534	0.848	0.952
7	1.134	0.419	0.204	5.204	0.076	1.924	1.182	0.118	1.882	2.704	0.833	0.959
8	1.061	0.373	0.388	5.306	0.136	1.864	1.099	0.185	1.815	2.847	0.820	0.965
9	1.000	0.337	0.547	5.393	0.184	1.816	1.032	0.239	1.761	2.970	0.808	0.969
10	0.949	0.308	0.687	5.469	0.223	1.777	0.975	0.284	1.716	3.078	0.797	0.973
11	0.905	0.285	0.811	5.535	0.256	1.744	0.927	0.321	1.679	3.173	0.787	0.975
12	0.866	0.266	0.922	5.594	0.283	1.717	0.886	0.354	1.646	3.258	0.778	0.978
13	0.832	0.249	1.025	5.647	0.307	1.693	0.850	0.382	1.618	3.336	0.770	0.979
14	0.802	0.235	1.118	5.696	0.328	1.672	0.817	0.406	1.594	3.407	0.763	0.981
15	0.775	0.223	1.203	5.741	0.347	1.653	0.789	0.428	1.572	3.472	0.756	0.982
16	0.750	0.212	1.282	5.782	0.363	1.637	0.763	0.448	1.552	3.532	0.750	0.984
17	0.728	0.203	1.356	5.820	0.378	1.622	0.739	0.466	1.534	3.588	0.744	0.985
18	0.707	0.194	1.424	5.856	0.391	1.608	0.718	0.482	1.518	3.640	0.739	0.985
19	0.688	0.187	1.487	5.891	0.403	1.597	0.698	0.497	1.503	3.689	0.733	0.986
20	0.671	0.180	1.549	5.921	0.415	1.585	0.680	0.510	1.490	3.735	0.729	0.987
21	0.655	0.173	1.605	5.951	0.425	1.575	0.663	0.523	1.477	3.778	0.724	0.988
22	0.640	0.167	1.659	5.979	0.434	1.566	0.647	0.534	1.466	3.819	0.720	0.988
23	0.626	0.162	1.710	6.006	0.443	1.557	0.633	0.545	1.455	3.858	0.716	0.989
24	0.612	0.157	1.759	6.031	0.451	1.548	0.619	0.555	1.445	3.895	0.712	0.989
25	0.600	0.153	1.806	6.056	0.459	1.541	0.606	0.565	1.435	3.931	0.708	0.990

Appendix P
C = 0 Sampling Plan
Adapted from
Zero Acceptance Number Sampling Plans, 5th Edition (used with permission)

Lot size	.010	.015	.025	.040	.065	.10	.15	.25	.40	.65	1.0	1.5	2.5	4.0	6.5	10.0
1–8	A	A	A	A	A	A	A	A	A	A	A	A	5	3	3	3
9–15	A	A	A	A	A	A	A	A	A	A	13	8	5	3	3	3
16–25	A	A	A	A	A	A	A	A	A	20	13	8	5	3	3	3
26–50	A	A	A	A	A	A	A	A	32	20	13	8	7	7	5	3
51–90	A	A	A	A	A	A	80	50	32	20	13	13	11	8	5	4
91–150	A	A	A	A	A	125	80	50	32	20	19	19	11	9	6	5
151–280	A	A	A	A	200	125	80	50	32	29	29	19	13	10	7	6
281–500	A	A	A	315	200	125	80	50	48	47	29	21	16	11	9	7
501–1200	A	800	500	315	200	125	80	75	73	47	34	27	19	15	11	8
1201–3200	1250	800	500	315	200	125	120	116	73	53	42	35	23	18	13	9
3201–10,000	1250	800	500	315	200	192	189	116	86	68	50	38	29	22	15	9
10,001–35,000	1250	800	500	315	300	294	189	135	108	77	60	46	35	29	15	9
35,001–150,000	1250	800	500	490	476	294	218	170	123	96	74	56	40	29	15	9
150,001–500,000	1250	800	750	715	476	345	270	200	156	119	90	64	40	29	15	9
500,001 and over	1250	1200	1112	715	556	435	303	244	189	143	102	64	40	29	15	9

"A" indicates that all parts must be inspected.

255

Appendix Q
Fall-Out Rates

C_{pk}	$Z(\sigma)$	One-tail $p(x)$	One-tail ppm	Two-tail $p(x)$	Two-tail ppm
0.00	0.00	0.500000	500000.0	1.000000	1000000.0
0.08	0.25	0.401294	401293.7	0.802587	802587.3
0.17	0.50	0.308538	308537.5	0.617075	617075.1
0.25	0.75	0.226627	226627.4	0.453255	453254.7
0.33	1.00	0.158655	158655.3	0.317311	317310.5
0.42	1.25	0.105650	105649.8	0.211300	211299.5
0.50	1.50	0.066807	66807.2	0.133614	133614.4
0.58	1.75	0.040059	40059.2	0.080118	80118.3
0.67	2.00	0.022750	22750.1	0.045500	45500.3
0.75	2.25	0.012224	12224.5	0.024449	24448.9
0.83	2.50	0.006210	6209.7	0.012419	12419.3
0.92	2.75	0.002980	2979.8	0.005960	5959.5
1.00	3.00	0.001350	1349.9	0.002700	2699.8
1.08	3.25	0.000577	577.0	0.001154	1154.1
1.17	3.50	0.000233	232.6	0.000465	465.3
1.25	3.75	0.000088	88.4	0.000177	176.8
1.33	4.00	0.000032	31.7	0.000063	63.3
1.42	4.25	0.000011	10.7	0.000021	21.4
1.50	4.50	0.000003	3.4	0.000007	6.8
1.58	4.75	0.000001	1.02	0.000002	2.03
1.67	5.00	0.000000	0.29	0.000001	0.57
1.75	5.25	0.000000	0.08	0.000000	0.15
1.83	5.50	0.000000	0.019	0.000000	0.038
1.92	5.75	0.000000	0.004	0.000000	0.009
2.00	6.00	0.000000	0.001	0.000000	0.002

Bibliography

Besterfield, Dale H. 2009. *Quality Control*. 8th ed. Columbus, OH: Pearson.

Burr, Irving W. 1974. *Applied Statistical Methods*. New York: Academic Press.

———. 1976. *Statistical Quality Control Methods*. New York: Marcel Dekker.

Dovich, Robert A. 1992. *Quality Engineering Statistics*. Milwaukee: ASQ Quality Press.

———. 1990. *Reliability Statistics*. Milwaukee: ASQ Quality Press.

Duncan, Archeson A. 1986. *Quality Control and Industrial Statistics*. Homewood, IL: Irwin.

Gojanovic, Tony. 2007. "Zero Defect Sampling." *Quality Progress*, November.

Grant, Eugene L., and Richard S. Leavenworth. 1980. *Statistical Quality Control*. New York: McGraw-Hill.

Hampel, Frank R. 1971. "A General Qualitative Definition of Robustness." *Annals of Mathematical Statistics* 42 (6): 1887–96.

———. 1974. "The Influence Curve and Its Role in Robust Estimation." *Journal of the American Statistical Association* 69 (346): 383–93.

Hansen, Bertrand L. 1963. *Quality Control: Theory and Applications*. Englewood Cliffs, NJ: Prentice Hall.

Hicks, Charles R. 1982. *Fundamental Concepts in the Design of Experiments*. New York: Holt, Rinehart and Winston.

Iglewicz, Boris, and David Caster Hoaglin. 1993. *How to Detect and Handle Outliers*. Milwaukee: ASQ Quality Press.

Juran, Joseph M., and Frank M. Gryna. 1988. *Juran's Quality Control Handbook*. 4th ed. New York: McGraw Hill.

Natrella, Mary G. 1966. *Experimental Statistics*. National Bureau of Standards Handbook 91. Washington, D.C.: U.S. Department of Commerce.

Nelson, Lloyd S. 1989. "Upper Confidence Limits on Average Number of Occurances." *Journal of Quality Technology* 21 (1) (January): 71–72

Snedecor, George W., and William G. Cochran. 1989. *Statistical Methods*. Ames, IA: Iowa State University Press.

Squeglia, Nicholas K. 2008. *Zero Acceptance Number Sampling Plans*. 5th ed. Milwaukee: ASQ Quality Press.

Stephens, Kenneth S. 2001. *The Handbook of Applied Acceptance Sampling: Plans, Procedures, and Principles*. Milwaukee: ASQ Quality Press.

Triola, Mario F. 2010. *Elementary Statistics*. 11th ed. New York: Addison-Wesley.

Walfish, Steven. 2006. "A Review of Statistical Outlier Methods." *Pharmaceutical Technology* (November).

Weingarten, Harry. 1982. "Confidence Intervals for the Percent Nonconforming Based on Variables Data." *Journal of Quality Technology* 14 (4) (October): 207–10.

Wise, Stephen A., and Douglas C. Fair. 1998. *Innovative Control Charting*. Milwaukee: ASQ Quality Press.

WEBSITES

American Society for Quality home page. http://asq.org/index.aspx

Elsmar Cove Discussion Forums. http://elsmar.com/Forums/index.php

International Seed Testing Association—Checking Data for Outliers, 2. 3σ Edit Rule and Hampel's Method. http://www.seedtest.org/upload/cms/user/presentation2Remund2.pdf

National Institute of Standards and Technology—Engineering Statistics Handbook. http://www.itl.nist.gov/div898/handbook/index.htm

QI Macros—Excel SPC software. http://www.qimacros.com/Macros.html

Wikipedia—One-Way Analysis of Variance. http://en.wikipedia.org/wiki/One-way_ANOVA

Wikipedia—Two-Way Analysis of Variance. http://en.wikipedia.org/wiki/Two-way_analysis_of_variance

Index

Ask a Librarian

Did you know?

- The ASQ Quality Information Center contains a wealth of knowledge and information available to ASQ members and non-members

- A librarian is available to answer research requests using ASQ's ever-expanding library of relevant, credible quality resources, including journals, conference proceedings, case studies and Quality Press publications

- ASQ members receive free internal information searches and reduced rates for article purchases

- You can also contact the Quality Information Center to request permission to reuse or reprint ASQ copyrighted material, including journal articles and book excerpts

- For more information or to submit a question, visit **http://asq.org/knowledge-center/ ask-a-librarian-index**

Visit www.asq.org/qic for more information.

ASQ
The Global Voice of Quality™

TRAINING   CERTIFICATION   CONFERENCES   MEMBERSHIP   **PUBLICATIONS**

Belong to the Quality Community!

Established in 1946, ASQ is a global community of quality experts in all fields and industries. ASQ is dedicated to the promotion and advancement of quality tools, principles, and practices in the workplace and in the community.

The Society also serves as an advocate for quality. Its members have informed and advised the U.S. Congress, government agencies, state legislatures, and other groups and individuals worldwide on quality-related topics.

Vision

By making quality a global priority, an organizational imperative, and a personal ethic, ASQ becomes the community of choice for everyone who seeks quality technology, concepts, or tools to improve themselves and their world.

ASQ is...

- More than 90,000 individuals and 700 companies in more than 100 countries

- The world's largest organization dedicated to promoting quality

- A community of professionals striving to bring quality to their work and their lives

- The administrator of the Malcolm Baldrige National Quality Award

- A supporter of quality in all sectors including manufacturing, service, healthcare, government, and education

- YOU

Visit www.asq.org for more information.

TRAINING   CERTIFICATION   CONFERENCES   MEMBERSHIP   **PUBLICATIONS**

ASQ Membership

Research shows that people who join associations experience increased job satisfaction, earn more, and are generally happier*. ASQ membership can help you achieve this while providing the tools you need to be successful in your industry and to distinguish yourself from your competition. So why wouldn't you want to be a part of ASQ?

Networking

Have the opportunity to meet, communicate, and collaborate with your peers within the quality community through conferences and local ASQ section meetings, ASQ forums or divisions, ASQ Communities of Quality discussion boards, and more.

Professional Development

Access a wide variety of professional development tools such as books, training, and certifications at a discounted price. Also, ASQ certifications and the ASQ Career Center help enhance your quality knowledge and take your career to the next level.

Solutions

Find answers to all your quality problems, big and small, with ASQ's Knowledge Center, mentoring program, various e-newsletters, *Quality Progress* magazine, and industry-specific products.

Access to Information

Learn classic and current quality principles and theories in ASQ's Quality Information Center (QIC), *ASQ Weekly* e-newsletter, and product offerings.

Advocacy Programs

ASQ helps create a better community, government, and world through initiatives that include social responsibility, Washington advocacy, and Community Good Works.

Visit www.asq.org/membership for more information on ASQ membership.

*2008, The William E. Smith Institute for Association Research

Notes

Notes

Notes

Notes